The R.A.M.S. Library of Alchemy

Volume 13

The Turba Philosophorum

by Arisleus

Also includes

Revelation of the True Chemical Wisdom

by Friederich Gualdus

R.A.M.S. Publishing Company

The Turba Philosophorum

by Arisleus

Also includes

Revelation of the True Chemical Wisdom

by Friederich Gualdus

Produced by

Restorers of Alchemical Manuscripts Society
1989

R.A.M.S. Publishing Company

R.A.M.S. Publishing Company
117 Rutherford Lane
Stuarts Draft VA 24477

First Edition 2015

ISBN-13 **978-1508856351**
ISBN-10 **1508856354**

Image Processing by Philip N. Wheeler

This book is sold for informational purposes only. Neither the publisher nor the editor shall be held accountable for the use or misuse of the information in this book.

Printed in the United States of America

Table of Contents

Dedicated to Hans W. Nintzel,

American Alchemist

and

Founder of the

Restorers of Alchemical Manuscripts Society

(R.A.M.S.)

Disclaimer

Liability: The publisher does not warrant or assume any legal liability or responsibility for the accuracy, completeness, or usefulness of any information, apparatus, product, or process disclosed. The publisher makes no representation as to the accuracy or completeness of the contents of this book and specifically disclaims any implied warranty of merchantability or fitness for a particular purpose. No warranty may be created or extended by written sales materials or sales representatives. You should obtain professional consultation where appropriate. The publisher shall not be liable for any loss of profit or other commercial or personal damages, including but not limited to special, incidental, consequential, or other damages.

Introduction

Philip N. Wheeler

The Turba Philosophorum, or Assembly of the Alchemical Philosophers, is attributed to Arisleus. It is one of the earliest Alchemical texts, believed to be from the 12th Century. The Turba Philosophorum was often quoted in later Alchemical texts.

Also included in this volume: *Revelation of the True Chemical Wisdom* by Friederich Gualdus, which includes a Forward by Hans W. Nintzel.

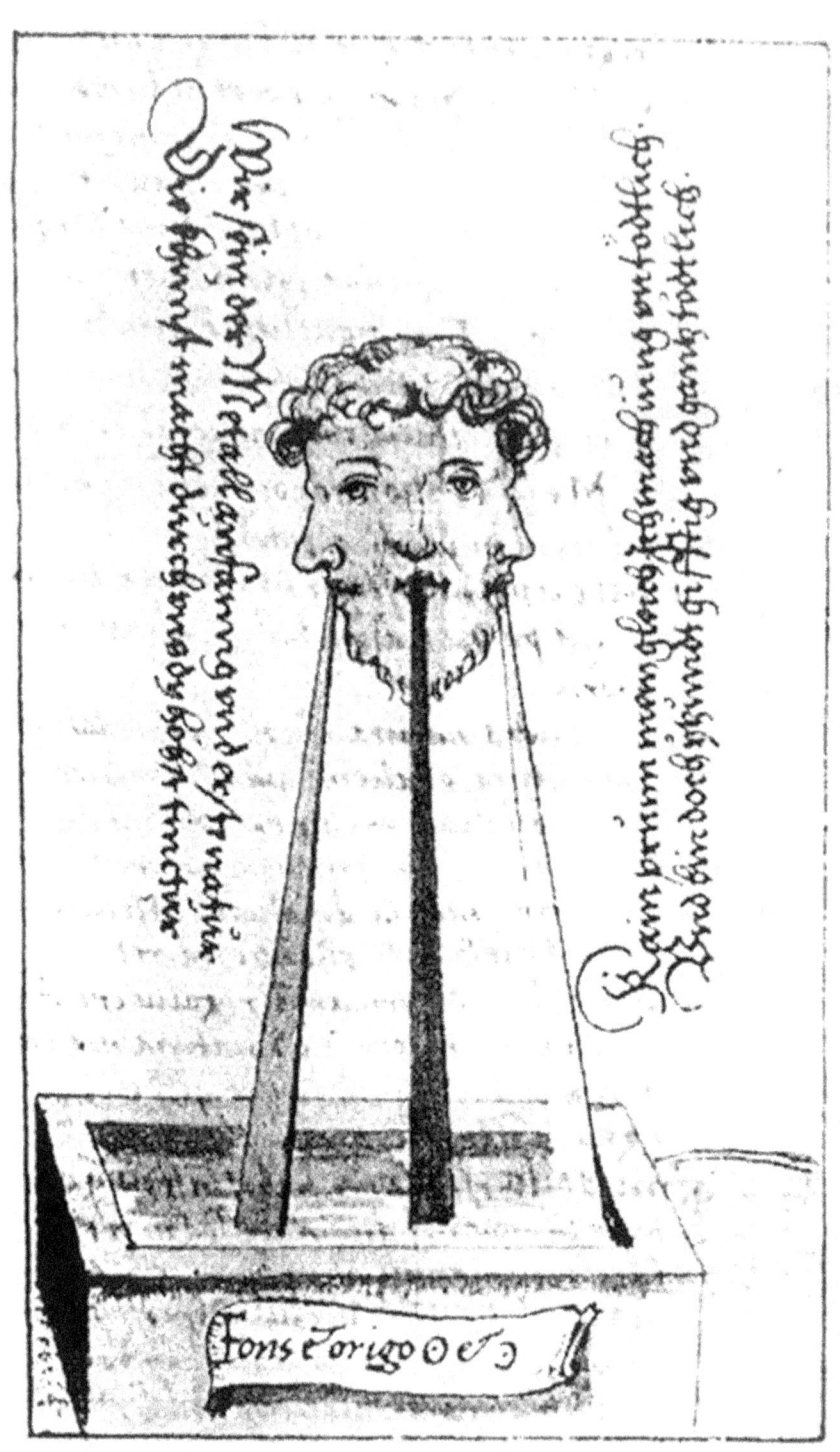

fons corigo ⊙ ☿

The Epistle of Arisleus, prefixed to the Words of the Sages, concerning the Purport of this Book, for the Benefit of Posterity, and the same being as here follows:

Arisleus, begotten of Pythagoras, a disciple of the disciples by the grace of thrice great Hermes, learning from the seat of knowledge, unto all who come after wisheth health and mercy. I testify that my master, Pythagoras, the Italian, master of the wise and chief of the Prophets, had a greater gift of God and of Wisdom than was granted to any one after Hermes. Therefore he had a mind to assemble his disciples, who were now greatly increased, and had been constituted the chief persons throughout all regions for the discussion of this most precious Art, that their words might be a foundation for posterity. He then commanded Iximidrus, of highest council, to be the first speaker, who said:

The First Dictum

Iximidrus Saith: I testify that the beginning of all things is a Certain Nature, which is perpetual, coequalling all things, and that the visible natures, with their births and decay, are times wherein the ends to which that nature brings them are beheld and summoned. Now, I instruct you that the stars are igneous, and are kept within bounds by the air. If the humidity and density of the air did not exist to separate the flames of the sun from living things, then the Sun would consume all creatures. But God has provided the separating air, lest that which He has created should be burnt up. Do you not: observe that the Sun when it rises in the heaven overcomes the air by its heat, and that the warmth penetrates from the upper to the lower parts of the air? If, then, the air did not presently breathe forth those winds whereby creatures are generated, the Sun by its heat would certainly destroy all that lives. But the Sun is kept in check by the air, which thus conquers because it unites the heat of the Sun to its own heat, and the humidity of water to its own humidity. Have you not remarked how tenuous water is drawn up into the air by the action of the heat of the Sun, which thus helps the water against itself? If the water did not nourish the air by such tenuous moisture, assuredly the Sun would overcome the air. The fire, therefore, extracts moisture from the water, by means of which the air conquers the fire itself. Thus, fire and water are enemies between which there is no consanguinity, for the fire is hot and dry, but the water is cold and moist. The air, which is warm and moist, joins these together by its concording medium; between the humidity of water and the heat of fire the air is thus placed to establish peace. rind look ye all how there shall arise a spirit from the tenuous vapour of the air, because the heat being joined to the humour, there necessarily issues something tenuous, which will become a wind. For the heat of the Sun extracts something tenuous out of the air, which also becomes spirit and life to all creatures. All this, however, is disposed in such manner by the will of God, and a coruscation appears when the heat of the Sun touches and breaks up a cloud.

The Turba saith: Well hast thou described the fire, even as thou knowest concerning it, and thou hast believed the word of thy brother.

The Second Dictum.

Exumedrus saith: I do magnify the air according to the mighty speech of Iximidrus, for the work is improved thereby. The air is inspissated, and it is also made thin; it grows warm and becomes cold. The inspissation thereof takes place when it is divided in heaven by the elongation of the Sun; its rarefaction is when, by the exaltation of the Sun in heaven, the air becomes warm and is rarefied. It is comparable with the complexion of Spring, in the distinction of time, which is neither warm nor cold. For according to the mutation of the constituted disposition with the altering distinctions of the soul, so is Winter altered. The air, therefore, is inspissated when the Sun is removed from it, and then cold supervenes upon men.

Whereat the Turba said: Excellently hast thou described the air, and given account of what thou knowest to be therein.

The Third Dictum.

Anaxagoras saith: I make known that the beginning of all those things which God hath created is weight and proportion, for weight rules all things, and the weight and spissitude of the earth is manifest in proportion; but weight is not found except in body. And know, all ye Turba, that the spissitude of the four elements reposes in the earth; for the spissitude of fire falls into air, the spissitude of air, together with the spissitude received from the fire, falls into water; the spissitude also of water, increased by the spissitude of fire and air, reposes in earth. Have you not observed how the spissitude of the four elements is conjoined in earth! The same, therefore, is more inspissated than all.

Then saith the Turba: Thou hast well spoken. Verily the earth is more inspissated than are the rest. Which, therefore, is the most rare of the four elements and is most worthy to possess the rarity of these four?

He answereth: Fire is the most rare among all, and thereunto cometh what is rare of these four. But air is less rare than fire, because it is warm and moist, while fire is warm and dry; now that which is warm and dry is more rare than the warm and moist. They say unto him: The which element is of less rarity than air!

He answereth: Water, since cold and moisture inhere therein, and every cold humid is of less rarity than a warm humid.

Then do they say unto him: Thou hast spoken truly. What, therefore, is of less rarity than water?
He answereth: Earth, because it is cold and dry, and that which is cold and dry is of less rarity than that which is cold and moist.

Pythagoras saith: Well have ye provided, O Sons of the Doctrine, the description of these four natures, out of which God hath created all things. Blessed, therefore, is he who comprehends what ye have declared, for from the apex of the world he shall not find an intention greater than his own! Let us, therefore, make perfect our discourse.

They reply: Direct everyone to take up our speech in turn. Speak thou, O Pandolfus!

The Fourth Dictum.

But Pandolfus saith: I signify to posterity that air is a tenuous matter of water, and that it is not: separated from it. It remains above the dry earth, to wit, the air hidden in the water, which is under the earth. If this air did not exist, the earth would not remain above the humid water. They answer: Thou hast said well; complete, therefore, thy speech.

But he continueth: The air which is hidden in the water under the earth is that which sustains the earth, lest it should be plunged into the said water; and it, moreover, prevents the earth from being overflowed by that water. The province of the air is, therefore, to fill up and to make separation between diverse things, that is to say, water and earth, and it is constituted a peacemaker between hostile things, namely, water and fire, dividing these, lest they destroy one another.

The Turba saith: If you gave an illustration hereof, it would be clearer to those who do not understand. He answereth: An egg is an illustration, for therein four things are conjoined; the visible cortex or shell represents the earth, and the albumen, for white part, is the water. But a very thin inner cortex is joined to the outer cortex, representing, as I have signified to you, the separating medium between earth and water, namely, that air which divides the earth from the water. The yolk also of the egg represents fire; the cortex which contains the yolk corresponds to that other air which separates the water from the fire. But they are both one and the same air, namely, that which separates things frigid, the earth from the water, and that which separates the water from the fire. But the lower air is thicker than the upper air, and the upper air is more rare and subtle, being nearer to the fire than the lower air. In the egg, therefore, are four things- earth, water, air, and fire. But the point of the Sun, these four excepted, is in the centre of the yolk, and this is the chicken. Consequently, all philosophers in this most excellent art have described the egg as an example, which same thing they have set over their work.

The Fifth Dictum.

Arisleus saith: Know that the earth is a hill and not a plain, for which reason the Sun does not ascend over all the zones of the earth in a single hour; but if it were flat, the sun would rise in a moment over the whole earth.

Parmenides saith: Thou hast spoken briefly, O Arisleus!

He answereth: Is there anything the Master has left us which bears witness otherwise? Yet I testify that God is one, having never engendered or been begotten, and that the head of all things after Him is earth and fire, because fire is tenuous and light, and it rules all things on earth, but the earth, being ponderous and gross, sustains all things which are ruled by fire.

The Sixth Dictum.

Lucas saith: You speak only about four natures; and each one of you observes something concerning these. Now, I testify unto you that all things which God hath created are from these four natures, and the things which have been created out of them return into them, In these living creatures are generated and die, and all things take place as God hath predestinated.

Democritus, the disciple of Lucas, answereth: Thou hast well spoken, O Lucas, when dealing with the four natures!

Then saith Arisleus: O Democritus, since thy knowledge was derived from Lucas, it is presumption to speak among those who are well acquainted with thy master!

Lucas answereth: albeit Democritus received from me the science of natural things, that knowledge was derived from the philosophers of the Indies and from the Babylonians; I think he surpasses those of his own age in this learning.

The Turba answereth: When he attains to that age he
will give no small satisfaction, but being in his
youth he should keep silence.

The Seventh Dictum.

Lucusta saith: All those creatures which have been described by Lucas are two only, of which one is neither known nor expressed, except by piety, for it is not seen or felt.

Pythagoras saith: Thou hast entered upon a subject which, if completed, thou wilt describe subtly. State, therefore, what is this thing which is neither felt, seen, nor known.

Then he: It is that which is not known, because in this world it is discerned by reason without the clients thereof, which are sight, hearing, taste, smell, and touch. O Crowd of the Philosophers, know you not that it Is only sight which can distinguish white from black, and hearing only which can discriminate between a good and bad word! Similarly, a wholesome odour cannot be separated by reason from one which is fetid, except through the sense of smell, nor can sweetness be discriminated from bitterness save by means of taste, nor smooth from rough unless by touch.

The Turba answereth: Thou hast well spoken, yet hast thou omitted to treat of that particular thing which is not known, or described, except by reason and piety.

Saith he: Are ye then in such haste! Know that the creature which is cognised in none of these five ways is a sublime creature, and, as such, is neither seen nor felt, but is perceived by reason alone, of which reason Nature confesses that God is a partaker.

They answer: Thou hast spoken truly and excellently. And he: I will now give a further explanation. Know that this creature, that is to say, the world, hath a light, which is the Sun, and the same is more subtle than all other natures, which light is so ordered that living beings may attain to vision. But if this subtle light were removed, they would become darkened, seeing nothing, except the light of the moon, or of the stars, or of fire, all which are derived from the light of the Sun, which causes all creatures to give light. For this God has appointed the Sun to be the light of the world, by reason of the attenuated nature of the Sun. And know that the sublime creature before mentioned has no need of the light of this Sun, because the Sun is beneath that creature, which is more subtle and more lucid. This light, which is more lucid than the light of the Sun, they have taken from the light of God, which is more subtle than their light. Know also that the created world is composed of two dense things and two rare things, but nothing of the dense is in the sublime creature. Consequently the Sun is rarer than all inferior creatures.

The Turba answereth: Thou hast excellently described what thou hast related. And if, good Master, thou shalt utter anything whereby our hearts may be vivified, which now are mortified by folly, thou wilt confer upon us a great boon!

The Eighth Dictum.

Pythagoras saith: I affirm that God existed before all things, and with Him was nothing, as He was at first. But know, all ye Philosophers, that I declare this in order that I may fortify your opinion concerning these four elements and arcana, as well as in the sciences thereof, at which no one can arrive save by the will of God. Understand, that when God was alone, He created four things- fire, air, water, and earth, out of which things He afterwards created all others, both the sublime and the inferior, because He predestinated from the beginning that all creatures extracted from water should multiply and increase, that they might dwell in the world and perform His judgments therein. Consequently, before all, He created the four elements, out of which He afterwards created what He willed, that is to say, diverse creatures, some of which were produced from a single element.

The Turba saith: Which are these, O Master!
And he: They are the angels, whom He created out of fire.

But the Turba: Which, then, are created out of two?

And he: Out of the elements of fire and air are the sun, moon, and stars composed. Hence the angels are more lucid than the sun, moon, and stars, because they are created from one substance, which is less dense than two, while the sun and the stars are created from a composition of fire and air.

The Turba saith: And what concerning the creation of Heaven?

Then he: God created the Heaven out of water and air, whence this is also composed of two, namely, the second of the rarer things, which is air, and the second of the denser things, which is water. And they: Master, continue thy discourse concerning these three, and rejoice our hearts with thy sayings, which are life to the dead.

But the other answereth: I notify to you that God hath further made creatures out of three and out of four; out of three are created flying things, beasts, and vegetables; some of these are created out of water, air, and earth, some out of fire, air, and earth.

But the Turba saith: Distinguish these divers creatures one from another.

And he: Beasts are created out of fire, air, and earth; dying things out of fire, air, and water, because flying things, and all among vegetables which have a spirit, are created out of water, while all brute animals are from earth, air, and fire. Yet in vegetables there is no fire, for they are created out of earth, water, and air.

Whereat the Turba saith: Let us assume that a fire, with your reverence's pardon, does reside in vegetables.

And he: Ye have spoken the truth, and I affirm that they contain fire.

And they: Whence is that fire?

He answereth: Out of the heat of the air which is concealed therein; for I have signified that a thin fire is present in the air, but the elementary fire concerning which you were in doubt is not produced, except in things which have spirit and soul. But out of four elements our father Adam and his sons were created, that is, of fire, air, water, and likewise earth. Understand, all ye that are wise, how everything which God hath created out of one essence dies not until the Day of Judgment. The definition of death is the disjunction of the composite, but there is no disjunction of that which is simple, for

it is one. Death consists in the separation of the
soul from the body, because anything formed out of
two, three, or four components must disintegrate,
and this is death. Understand, further, that no
complex substance which lacks fire eats, drinks, or
sleeps, because in all things which have a spirit
fire is that which eats.

The Turba answereth: How is it, Master, that the
angels, being created of fire, do not eat, seeing
thou assertest that fire is that which eats!
And he: Hence ye doubt, each having his opinion, and
ye are become opponents, but if ye truly knew the
elements, ye would not deny these things. I agree
with all whose judgment it is that simple fire eats
not, but thick fire. The angels, therefore, are not
created out of thick fire, but out of the thinnest
of very thin fire; being created, then, of that
which is most simple and exceedingly thin, they
neither eat, drink, nor sleep.

And the Turba: Master, our faculties are able to
perceive, for by God's assistance we have exhausted
thy sayings, but our faculties of hearing and of
sight are unable to carry such great things. May God
reward thee for the sake of thy disciples, since it
is with the object of instructing future generations
that thou hast summoned us together from our

countries, the recompense of which thou wilt not fail to receive from the Judge to come.

Arisleus saith: Seeing that thou hast gathered us together for the advantage of posterity, I think that no explanations will be more useful than definitions of those four elements which thou hast taught us to attain.

And he: None of you are, I suppose, ignorant that all the Wise have propounded definitions in God.

The Turba answereth: Should your disciples pass over anything, it becomes you, O Master, to avoid omissions for the sake of future generations.

And he: If it please you, I will begin the disposition here, since envious men in their books have separated that, or otherwise I will put it at the end of the book.

Whereat the Turba saith: Place it where you think it will be dearest for future generations.

And he: I will place it where it will not be recognised by the foolish, nor ignored by the Sons of the Doctrine, for it is the key, the perfection and the end.

The Ninth Dictum.

Eximenus saith: God hath created all things by his word, having said unto them: Be, and they were made, with the four other elements, earth, water, air, and tire, which He coagulated, and things contrary were commingled, for we see that fire is hostile to water, water hostile to fire, and both are hostile to earth and air. Yet God hath united them peacefully, so that they love one another. Out of these four elements, therefore, are all things created: heaven and the throne thereof; the angels; the sun, moon and stars; earth and sea, with all things that are in the sea, which indeed are various, and not alike, for their natures have been made diverse by God, and also the creations. But the diversity is more than I have stated; each of these natures is of diverse nature, and by a legion of diversities is the nature of each diverse. Now this diversity subsists in all creatures, because they were created out of diverse elements. Had they been created out of one element, they would have been agreeing natures. But diverse elements being here mingled, they lose their own natures, because the dry being mixed with the humid and the cold combined with the hot, become neither cold nor hot; so also the humid being mixed with the dry becomes neither dry nor humid. But when the four elements are commingled, they agree, and thence proceed creatures which never attain to perfection, except they be left by night to putrefy and become visibly corrupt. God further completed his creation by means of increase, food, life, and government. Sons of the Doctrine, not without purpose have I described to you the disposition of these four elements, for in them is a secret arcanum; two of them are perceptible to the sense of touch and vision, and of these the operation and virtue are well known. These are earth and water. But there are two other elements which are neither visible nor tangible, which yield naught, whereof the place is never seen, nor are their operations and force known, save in the former elements, namely, earth and water; now when the four elements are not commingled, no desire of men is accomplished. But being mixed, departing from their own natures, they become another thing. Over these let us meditate very carefully.

And the Turba: Master, if you speak, we will give heed to Your words.

Then he: I have now discoursed, and that well. I will speak only useful words which ye will follow as spoken. Know, all present, that no true tincture is made except from our copper. Do not therefore, exhaust your brains and your money, lest ye fill your hearts with sorrow. I will give you a fundamental axiom, that unless you turn the aforesaid copper into white, and make visible coins and then afterwards again turn it into redness, until a Tincture: results, verily, ye accomplish nothing. Burn therefore the copper, break it up, deprive it of its blackness by cooking, imbuing, and washing, until the same becomes white. Then rule it.

The Tenth Dictum.

Arisleus saith: Know that the key of this work is the art of Coins. Take, therefore, the body which I have shewn to you and reduce it to thin tablets. Next immerse the said tablets in the Water of our Sea, which is permanent Water, and, after it is covered, set it over a gentle fire until the tablets are melted and become waters or Etheliae, which are one and the same thing. Mix, cook, and simmer in a gentle fire until Brodium is produced, like to Saginatum. Then stir in its water of Etheliae until it be coagulated, and the coins become variegated, which we call the Flower of Salt. Cook it, therefore, until it be deprived of blackness, and the whiteness appear. Then rub it, mix with the Gum of Gold, and cook until it becomes red Etheliae. Use patience in pounding lest you become weary. Imbue the Ethelia with its own water, which has preceded from it, which also is Permanent Water, until the same becomes red. This, then, is Burnt Copper, which is the Leaven of Gold and the Flower thereof. Cook the same with Permanent Water, which is always with it, until the water be dried up. Continue the operation until all the water is consumed, and it becomes a most subtle powder.

The Eleventh Dictum.

Parmenides saith: Ye must know that envious men have dealt voluminously with several waters, brodiums, stones, and metals, seeking to deceive all you who aspire after knowledge. Leave, therefore, all these, and make the white red, out of this our copper, taking copper and lead, letting these stand for the grease, or blackness, and tin for the liquefaction. Know ye, further, that unless ye rule the Nature of Truth, and harmonize well together its complexions and compositions, the consanguineous with the consanguineous, and the first with the first, ye act improperly and effect nothing, because natures will meet their natures, follow them, and rejoice. For in them they putrefy and are generated, because Nature is ruled by Nature, which destroys it, turns it into dust, reduces to nothing, and finally herself renews it, repeats, and frequently produces the same. Therefore look in books, that ye may know the Nature of Truth, what putrefies it and what renews, what savour it possesses, what neighbours it naturally has, and how they love each other, how also after love enmity and corruption intervene, and how these natures should be united one to another and made at peace, until they become gentle in the fire in similar fashion. Having, therefore, noticed the facts in this Art, set your hands to the work. If indeed, ye know not the Natures of Truth, do not approach the work, since there will follow nothing but harm, disaster, and sadness. Consider, therefore, the teaching of the Wise, how they have declared the whole work in this saying: Nature rejoices in Nature, and Nature contains Nature. In these words there is shewn forth unto you the whole work. Leave, therefore, manifold and superfluous things, and take quicksilver, coagulate in the body of Magnesia, in Kuhul, or in Sulphur which does not burn; make the same nature white, and place it upon our Copper, when it becomes white. And if ye cook still more, it becomes red, when if ye proceed to coction, it becomes gold. I tell you that it turns the sea itself into red and the colour of gold. Know ye also that gold is not turned into redness save by Permanent Water, because Nature rejoices in Nature.: Reduce, therefore, the same by means of cooking into a humour, until the hidden nature appear. If, therefore, it be manifested externally, seven times imbue the same with water, cooking, imbuing, and washing, until it become red. O those celestial natures, multiplying the natures of truth by the will of God! O that potent Nature, which overcame and conquered natures, and caused its natures to rejoice and be glad! This, therefore, is that special and spiritual nature to which the God thereof can give what fire cannot. Consequently, we glorify and magnify that [species], than which nothing is more precious in the true tincture, or the like in the smallest degree to be found. This is that truth which those investigating wisdom love. For when it is liquefied with bodies, the highest operation is effected. If ye knew the truth, what great thanks ye would give me! Learn, therefore, that while you are tingeing the cinders, you must destroy those that are mixed. For it overcomes those which are mixed, and changes them to its own colour. And as it visibly overcame the surface, even so it mastered the interior. And if one be volatile but the other endure the fire, either joined to the other endures the fire. Know also, that if the vapours have whitened the surfaces, they will certainly whiten the interiors. Know further, all ye seekers after Wisdom, that one matter overcomes four, and our Sulphur alone consumes all things.

The Turba answereth: Thou hast spoken excellently well, O Parmenides, but thou hast not demonstrated the disposition of the smoke to posterity, nor how the same is whitened!

The Twelfth Dictum.

Lucas saith: I will speak at this time, following the steps of the ancients. Know, therefore, all ye seekers after Wisdom, that this treatise is not from the beginning of the ruling! Take quicksilver, which is from the male, and coagulate according to custom. Observe that I am speaking to you in accordance with custom, because it has been already coagulated. Here, therefore, is not the beginning of the ruling, but I prescribe this method, namely, that you shall take the quicksilver from the male, and shall either impose upon iron, tin, or governed copper, and it will be whitened. White Magnesia is made in the same way, and the male is converted with it. But forasmuch as there is a certain affinity between the magnet and the iron, therefore our nature rejoices.) Take, then, the vapour which the Ancients commanded you to take, and cook the same with its own body until tin is produced. Wash away its blackness according to custom, and cleanse and roast at an equable fire until it be whitened. But every body is whitened with governed quicksilver, for Nature converts Nature. Take, therefore, Magnesia, Water of Alum, Water of Nitre, Water of the Sea, and Water of Iron; whiten with smoke: Whatsoever ye desire to be whitened is whitened with this smoke, because it is itself white, and whitens all things. Mix, therefore, the said smoke with its faeces until it be coagulated and become excessively white. Roast this white copper till it germinates of itself, since the Magnesia when whitened does not suffer the spirits to escape, or the shadow of copper to appear, because Nature contains Nature. Take, therefore, all ye Sons of the Doctrine, the white sulphureous nature, whiten with salt and dew, or with the Flower of White Salt, until it become excessively white. And know ye, that the Flower of White Salt is Ether from Ethelia. The same must be boiled for seven days, till it shall become like gleaming marble, for when it has reached this condition it is a very great Arcanum, seeing that Sulphur is mixed with Sulphur, whence an excellent work is accomplished, by reason of the affinity between them, because natures rejoice in meeting their own natures. Take, therefore, Mardek and whiten the same with Gadenbe, that is, wine and vinegar, and Permanent Water. Roast and coagulate until the whole does not liquefy in a fire stronger than its own, namely, the former fire. Cover the mouth of the vessel securely, but let it be associated with its neighbour, that it may kindle the whiteness thereof, and beware lest the fire blaze up, for in this case it becomes red prematurely, and this will profit you nothing, because in the beginning of the ruling you require the white. Afterwards coagulate the same until you attain the red. Let your fire be gentle in the whitening, until coagulation take place. Know that when it is coagulated we call it the Soul, and it is more quickly converted from nature into nature. This, therefore, is sufficient for those who deal with the Art of Coins, because one thing makes it but many operate therein. For ye need not a number of things, but one thing only, which in each and every grade of your work is changed into another nature.

The Turba saith: Master, if you speak as the Wise have spoken, and that briefly, they will follow you who do not wish to be wholly shut in with darkness.

The Thirteenth Dictum.

Pythagoras saith: We posit another government which is not from another root, but it differs in name. And know, all ye seekers after this Science and Wisdom, that whatsoever the envious may have enjoined in their books concerning the composition of natures which agree together, in savour there is only one, albeit to sight they are as diverse as possible. Know, also, that the thing which they have described in so many ways follows and attains its companion without fire, even as the magnet follows the iron, to which the said thing is not vainly compared, nor to a seed, nor to a matrix, for it is also like unto these. And this same thing, which follows its companion without fire, causes many colours to appear when embracing it, for this reason, that the said one thing enters into every regimen, and is found everywhere, being a stone, and also not a stone; common and precious; hidden and concealed, yet known by everyone; of one name and of many names, which is the Spume of the Moon. This stone, therefore, is not a stone, because it is more precious; without it Nature never operates anything; its name is one, yet we have called it by many names on account of the excellence of its nature.

The Turba answereth: O! Master! wilt thou not mention some of those names for the guidance of seekers?

And he: It is called White Ethelia, White Copper, and that which flies from the fire and alone whitens copper. Break up, therefore, the White Stone, and afterwards coagulate it with milk. Then pound the calx in the mortar, taking care that the humidity does not escape from the vessel; but coagulate it in the vessel until it shall become a cinder. Cook also with Spume of Luna and regulate. For ye shall find the stone broken, and already imbued with its own water. This, therefore, is the stone which we call by all names, which assimilates the work and drinks it, and is the stone out of which also all colours appear. Take, therefore, that same gum, which is from the scoriae, and mix with cinder of calx, which you have ruled, and with the faeces which you know, moistening with permanent water. Then look and see whether it has become a powder, but if not, roast in a fire stronger than the first fire, until it be pounded. Then imbue with permanent water, and the more the colours vary all the more suffer them to be heated. Know, moreover, that if you take white quicksilver, or the Spume of Luna, and do as ye are bidden, breaking up with a gentle fire, the same is coagulated, and becomes a stone. Out of this stone, therefore, when it is broken up, many colours will appear to you. But herein, if any ambiguity occur to you in our discourse, do as ye are bidden, ruling the same until a white and coruscating stone shall be produced, and so ye find your purpose.

The Fourteenth Dictum.

Acsubofen saith: Master, thou hast spoken without envy, even as became thee, and for the same may God reward thee!

Pythagoras saith: May God also deliver thee, Acsubofen, from envy!

Then he: Ye must know, O Assembly of the Wise, that sulphurs are contained in sulphurs, and humidity in humidity.

The Turba answereth: The envious, O Acsubofen, have uttered something like unto this! Tell us, therefore, what is this humidity?

And he: Humidity is a venom, and when venom penetrates a body, it tinges it with an invariable colour, and in no wise permits the soul to be separated from the body, because it is equal thereto. Concerning this, the envious have said: When one flies and the other pursues, then one seizes upon the other, and afterwards they no longer flee, because Nature has laid hold of its equal, after the manner of an enemy, and they destroy one another. For this reason, out of the sulphureous mixed sulphur is produced a most precious colour, which varies not, nor flees from the fire, when the soul enters into the interior of the body and holds the body together and tinges it. I will repeat my words in Tyrian dye. Take the Animal which is called Kenckel, since all its water is a Tyrian colour, and rule the same with a gentle fire, as is customary, until it shall become earth, in which there will be a little colour. But if you wish to obtain the Tyrian tincture, take the humidity which that thing has ejected, and place it therewith gradually in a vessel, adding that tincture whereof the colour was disagreeable to you. Then cook with that same marine water until it shall become dry. Afterwards moisten with that humour, dry gradually, and cease not to imbue it, to cook, and to dry, until it be imbued with all its humour. Then leave it for several days in its own vessel, Until the most precious Tyrian colour shall come out from it to the surface. Observe how I describe the regimen to you! Prepare it with the urine of boys, with water of the sea, and with permanent clean water, so that it may be tinged, and decoct with a gentle fire, until the blackness altogether shall depart from it, and it be easily pounded. Decoct, therefore, in its own humour until it clothe itself with a red colour. But if ye wish to bring it to the Tyrian colour, imbue the same with continual water, and mix, as ye know to be sufficient, according to the rule of sight; mix the same with permanent water sufficiently, and decoct until rust absorb the water. Then wash with the water of the sea which thou hast prepared, which is water of desiccated calx; cook until it imbibe its own moisture; and do this day by day. I tell you that a colour will thence appear to you the like of which the Tyrians have never made. And if ye wish that it should be a still more exalted colour, place the gum in the permanent water, with which ye shall dye it alternately, and afterwards desiccate in the sun. Then restore to the aforesaid water and the black Tyrian colour is intensified. But know that ye do not tinge the purple colour except by cold. Take, therefore, water which is of the nature of cold, and steep wool therein until it extract the force of the tincture from the water. Know also that the Philosophers have called the force which proceeds from that water the Flower. Seek, therefore, your intent in the said water; therein place what is in the vessel for days and nights, until it be clothed with a most precious Tyrian colour.

The Fifteenth Dictum.

Frictes saith: O all ye seekers after Wisdom, know that the foundation of this Art, on account of which many have perished, is one only. There is one thing which is stronger than all natures, and more sublime in the opinion of philosophers, whereas with fools it is more common than anything. But for us it is a thing which we reverence. Woe unto all ye fools! How ignorant are ye of this Art, for which ye would die if ye knew it! I swear to you that if kings were familiar with it, none of us would ever attain this thing. O how this nature changeth body into spirit! O how admirable is Nature, how she presides over all, and overcomes all!

Pythagoras saith: Name this Nature, O Frictes!

And he: It is a very sharp vinegar, which makes gold into sheer spirit, without which vinegar, neither whiteness, nor blackness, nor redness, nor rust can be made. And know ye that when it is mixed with the body, it is contained therein, and becomes one therewith; it turns the same into a spirit, and tinges with a spiritual and invariable tincture, which is indelible. Know, also, that if ye place the body over the fire without vinegar, it will be burnt

48

and corrupted. And know, further, that the first humour is cold. Be careful, therefore, of the fire, which is inimical to cold. Accordingly, the Wise have said: "Rule gently until the sulphur becomes incombustible." The Wise men have already shewn to those who possess reason the disposition of this Art, and the best point of their Art, which they mentioned, is, that a little of this sulphur burns a strong body. Accordingly they venerate it and name it in the beginning of their book, and the son of Adam thus described it. For this vinegar burns the body, converts it into a cinder, and also whitens the body, which, if ye cook well and deprive of blackness, is changed into a stone, so that it becomes a coin of most intense whiteness. Cook, therefore, the stone until it be disintegrated, and then dissolve and temper with water of the sea. Know also, that the beginning of the whole work is the whitening, to which succeeds the redness, finally the perfection of the work; but after this, by means of vinegar, and by the will of Gcd, there follows a complete perfection, Now, I have shewn to you, O disciples of this Turba, the disposition of the one thing, which is more perfect, more precious, and more honourable, than all natures, and I swear to you by God that I have searched for a long time in books so that I might arrive at the knowledge of this one thing, while I prayed also to God that he

would teach me what it is. My prayer was heard, He
shewed me clean water, whereby I knew pure vinegar,
and the more I did read books, the more was I
illuminated.

The Sixteenth Dictum.

Socrates saith: Know, O crowd of those that still remain of the Sons of the Doctrine, that no tincture can be produced without Lead, which possesses the required virtue. Have ye not seen how thrice-great Hermes infused the red into the body, and it was changed into an invariable colour? Know, therefore, that the first virtue is vinegar, and the second is the Lead of which the Wise have spoken, which if it be infused into all bodies, renders all unchangeable, and tinges them with an invariable colour. Take, therefore, Lead which is made out of the stone called Kuhul; let it be of the best quality, and let it be cooked till it becomes black. Then pound the same with Water of Nitre until it is thick like grease, and cook again in a very bright fire until the spissitude of the body is destroyed, the water being rejected. Kindle, therefore, above it until the stone becomes clean, abounding in precious metal, and exceedingly white. Pound it afterwards with dew and the Sun, and with sea and rain water for 31 days, for 10 days with salt water, and 10 days with fresh water, when ye shall find the same like to a metallic stone. Cook the same once more with water of nitre until it become tin by liquefaction. Again cook until it be deprived of

moisture, and become dry. But know that when it
becomes dry it drinks up what remains of its humour
swiftly, because it is burnt lead. Take care,
however, lest it be burnt. Thus we call it
incombustible sulphur. Pound the same with the
sharpest vinegar, and cook till it becomes thick,
taking care lest the vinegar be changed into smoke
and perish; continue this coction for 150 days. Now,
therefore, I have demonstrated the disposition of
the white lead, all which afterwards follows being
no more than women's work and child's play. Know,
also, that the arcanum of the work of gold proceeds
out of the male and the female, but I have shewn you
the male in the lead, while, in like manner, I have
discovered for you the female in orpiment. Mix,
therefore, the orpiment with the lead, for the
female rejoices in receiving the strength of the
male, because she is assisted by the male. But the
male receives a tingeing spirit from the female. Mix
them, therefore, together, place in a glass vessel,
and pound with Ethelia and very sharp vinegar; cook
for seven days, taking care lest the arcanum smoke
away, and leave throughout the night. But if ye wish
it to put on mud (colour), seeing that it is already
dry, again imbue with vinegar. Now, therefore, I
have notified to you the power of orpiment, which is
the woman by whom is accomplished the most great
arcanum. Do not shew these unto the evil, for they

will laugh. It is the Ethelia of vinegar which is placed in the preparation, by which things God perfects the work, whereby also spirits take possession of bodies, and they become spiritual.

The Seventeenth Dictum.

Zimon saith: O Turba of Philosophers and disciples, now hast thou spoken about making into white, but it yet remains to treat concerning the reddening! Know, all ye seekers after this Art, that unless ye whiten, ye cannot make red, because the two natures are nothing other than red and white. Whiten, therefore, the red, and redden the white! Know, also, that the year is divided into four seasons; the first season is of a frigid complexion, and this is Winter; the second is of the complexion of air, and this is Spring; then follows the third, which is summer, and is of the complexion of fire; lastly, there is the fourth, wherein fruits are matured, which is Autumn. In this manner, therefore, ye are to rule your natures, namely, to dissolve ill winter, to cook in spring, to coagulate in summer, and to gather and tinge the fruit in autumn. Having, therefore, given this example, rule the tingeing natures, but if ye err, blame no one save yourselves.

The Turba answereth: Thou hast treated the matter extremely well; add, therefore, another teaching of this kind for the sake of posterity.

And he: I will speak of making lead red. Take the copper which the Master ordered you to take at the beginning of his book, combine lead therewith, and cook it until it becomes thick; congeal also and desiccate until it becomes red. Here certainly is the Red Lead of which the wise spake; copper and lead become a precious stone; mix them equally, let gold be roasted with them, for this, if ye rule well, becomes a tingeing spirit in spirits. So when the male and the female are conjoined there is not produced a volatile wife, but a spiritual composite. From the composite turned into a red spirit is produced the beginning of the world. Behold this is the lead which we have called Red Lead, which is of our work, and without which nothing is effected!

The Eighteenth Dictum.

Mundus saith to the Turba: The seekers after this Art must know that the Philosophers in their books have described gum in many ways, but it is none other than permanent water, out of which our precious stone is generated. O how many are the seekers after this gum, and how few there are who find it! Know that this gum is not ameliorated except by gold alone. For there be very many who investigate these applications, and they find certain things, yet they cannot sustain the labours because they are diminished. But the applications which are made out of the gum and out of the honourable stone, which has already held the tincture, they sustain the labours, and are never diminished. Understand, therefore, my words, for I will explain unto you the applications of this gum, and the arcanum existing therein. Know ye that our gum is stronger than gold, and all those who know it do hold it more honourable than gold, yet gold we also honour, for without it the gum cannot be improved. Our gum, therefore, is for Philosophers more precious and more sublime than pearls, because out of gum with a little gold we buy much. Consequently, the Philosophers, when committing these things to writing that the same might not perish, have not set forth in their books the manifest disposition, lest every one should become acquainted therewith, and having become familiar to fools, the same would not sell it at a small price. Take, therefore, one part of the most intense white gum; one part of the urine of a white calf; one part of the gall of a fish; and one part of the body of gum, without which it cannot be improved; mix these portions and cook for forty days. When these things have been done, congeal by the heat of the sun till they are dried. Then cook the same, mixed with milk of ferment, until the milk fail; afterwards extract it, and until it become dry evaporate the moisture by heat. Then mix it with milk of the fig, and cook it till that moisture be dried up in the composite, which afterwards mix with milk of the root of grass, and again cook until it be dry. Then moisten it with rainwater, then sprinkle with water of dew, and cook until it be dried. Also imbue with permanent water, and desiccate until it become of the most intense dryness. Having done these things: mix the same with the gum which is equipped with all manner of colours, and cook strongly until the whole force of the water perish; and the entire body be deprived of its humidity, while ye imbue the same by cooking, until the dryness thereof be kindled. Then dismiss for forty days. Let it remain in that trituration or decocting until the spirit penetrate the body. For by this regimen the spirit is made corporeal, and the body is changed into a spirit. Observe the vessel, therefore, lest the composition fly and pass off in fumes. These things being accomplished, open the vessel, and ye will find that which ye purposed. This, therefore, is the arcanum of gum, which the Philosophers have concealed in their books.

The Nineteenth Dictum.

Dardaris saith: It is common knowledge that the Masters before us have described Permanent Water. Now, it behoves one who is introduced to this Art to attempt nothing till he is familiar with the power of this Permanent Water, and in commixture, contrition, and the whole regimen, it behoves us to use invariably this famous Permanent Water. He, therefore, who does not understand Permanent Water, and its indispensable regimen, may not enter into this Art, because nothing is effected without the Permanent Water. The force thereof is a spiritual blood, whence the Philosophers have called it Permanent Water, for, having pounded it with the body, as the Masters before me have explained to you, by the will of God it turns that body into spirit. For these, being mixed together and reduced to one, transform each other; the body incorporates the spirit, and the spirit incorporates the body into tinged spirit, like blood. And know ye, that whatsoever hath spirit the same hath blood also as well. Remember, therefore, this arcanum!

The Twentieth Dictum.

Belus saith: O disciples, ye have discoursed excellently!

Pythagoras answers: Seeing that they are philosophers, O Belus, why hast thou called them disciples?

He answereth: It is in honour of their Master, lest I should make them equal with him.

Then Pythagoras saith: Those who, in conjunction with us, have composed this book which is called the Turba, ought not to be termed disciples.
Then he: Master, they have frequently described Permanent Water, and the making of the White and the Red in many ways, albeit under many names; but in the modes after which they have conjoined weights, compositions, and regimens, they agree with the hidden truth. Behold, what is said concerning this despised thing! A report has gone abroad that the Hidden Glory of the Philosophers is a stone and not a stone, and that it is called by many names, lest the foolish should recognise it, Certain wise men have designated it after one fashion, namely, according to the place where it is generated; others

have adopted another, founded upon its colour, some
of whom have termed it the Green Stone; by other
some it is called the Stone of the most intense
Spirit of Brass, not to be mixed with bodies; by yet
others its description has been further varied,
because it is sold for coins by lapidaries who are
called saven; some have named it Spume of Luna; some
have distinguished it astronomically or
arithmetically; it has already received a thousand
titles, of which the best is: "That which is
produced out of metals." So also others have called
it the Heart of the Sun, and yet others have
declared it to be that which is brought forth out of
quicksilver with the milk of volatile things.

The Twenty-first Dictum.

Pandolfus saith: O Belus, thou hast said so much concerning the despised stone that thou hast left nothing to be added by thy brethren! Howsoever, I teach posterity that this despised stone is a permanent water, and know, all ye seekers after Wisdom, that permanent water is water of mundane life, because, verily, Philosophers have stated that Nature rejoices in Nature, Nature contains Nature, and Nature overcomes Nature. The Philosophers have constituted this short dictum the principle of the work for reasonable persons. And know ye that no body is more precious or purer than the Sun, and that no tingeing venom: is generated without the Sun and its shadow. He, therefore, who attempts to make the venom of the Philosophers without these, already errs, and has fallen into that pit wherein his sadness remains. But he who has tinged the venom of the wise out of the Sun and its shadow has arrived at the highest Arcanum. Know also that our coin when it becomes red, is called gold; he, therefore, who knows the hidden Cambar of the Philosophers, to him is the Arcanum already revealed.

The Turba answereth: Thou hast even now intelligibly described this stone, yet thou hast not narrated its regimen nor its composition. Return, therefore, to the description.

He saith: I direct you to take an occult and honourable arcanum, which is White Magnesia, and the same is mixed and pounded with wine, but take care not to make use of this except it be pure and clean; finally place it in its vessel, and pray God that He may grant you the sight of this very great stone. Then cook gradually, and, extracting, see if it has become a black stone, in which case ye have ruled excellently well. But rule it thus for the white, which is a great arcanum, until it becomes Kuhul, closed up with blackness, which blackness see that it does not remain longer than forty days. Pound the same, therefore, with its confections, which are the said flower of copper, gold of the Indies whose root is one, and a certain extract of an unguent, that is, of a crocus, that is, fixed exalted alum; cook the four, therefore, permanently for 40 or 42 days. After these days God will show you the principle(or beginning) of this stone, which is the stone Atitos, of which favoured sight of God there are many accounts. Cook strongly, and imbue with the gum that remains. And know ye that so often as ye imbue the cinder, so often must it be desiccated and again humectated, until its colour turns into that which ye desire. Now, therefore, will I complete that which I have begun, if God will look kindly on us. Know also that the perfection of the work of this precious stone is to rule it with the residue of the third part of the medicine, and to preserve the two other parts for imbuing and cooking alternately till the required colour appears. Let the fire be more intense than the former; let the matter be cerated, and when it is desiccated it coheres. Cook, therefore, the wax until it imbibes the gluten of gold, which being desiccated, imbue the rest of the work seven times until the other two thirds be finished, and true earth imbibe them all. Finally, place the same on a hot fire until the earth extract its flower and be satisfactory. Blessed are ye if ye understand! But, if not, I will repeat to you the perfection of the work. Take the clean white, which is a most great arcanum, wherein is the true tincture; imbue sand therewith, which sand is made out of the stone seven times imbued, until it drink up the whole, and close the mouth of the vessel effectually, as you have often been told. For that which ye seek of it by the favour of God, will appear to you, which is the stone of Tyrian colour. Now, therefore, I have fulfilled the truth, so do I conjure you by God and your sure Master, that you show not this great arcanum, and beware of the wicked!

The Twenty-Second Dictum.

Theophilus saith: Thou hast spoken intelligently and elegantly, and art held free from envy.

Saith the Turba: Let your discretion, therefore, explain to us what the instructing Pandolfus has stated, and be not envious.

Then he: O all ye seekers after this science, the arcanum of gold and the art of the coin is a dark vestment, and no one knows what the Philosophers have narrated in their books without frequent reading, experiments, and questionings of the Wise. For that which they have concealed is more sublime and obscure than it is possible to make known in words, and albeit some have dealt with it intelligibly and well, certain others have treated it obscurely; thus some are more lucid than others.

The Turba answereth: Thou hast truly spoken. And he: I announce to posterity that between boritis and copper there is an affinity, because the boritis of the Wise liquefies; the copper, and it changes as a fluxible water. Divide, therefore, the venom into two equal parts, with one of which liquefy the copper, but preserve the other to Pound and imbue the same, until it is drawn out into plates; cook again with the former part of the venom, cook two to seven in two; cook to seven in its own water for 42 days; finally, open the vessel, and ye shall find copper turned into quicksilver; wash the same by cooking until it be deprived of its blackness, and become as copper without a shadow. Lastly, cook it continuously until it be congealed. For when it is congealed it becomes a very great arcanum. Accordingly, the Philosophers have called this stone Boritis; cook, therefore, that coagulated stone until it becomes a matter like mucra. Then imbue it with the Permanent water which I directed you to reserve, that is to say, with the other portion, and cook it many times until its colours manifest. This, therefore, is the very great putrefaction which extracts (or contains in itself) the very great arcanum.

Saith the Turba: Return to thine exposition, O Theophilus!

And he: It is to be known that the same affinity which exists between the magnet and iron, also exists assuredly between copper and permanent water. If, therefore, ye rule copper and permanent water as I have directed, there will thence result the very great arcanum in the following fashion. Take white

Magnesia and quicksilver, mix with the male, and pound strongly by cooking, not with the hands, until the water become thin. But dividing this water into two parts, in the one part of the water cook it for eleven, otherwise, forty days, until there be a white flower, as the flower of salt in its splendour and coruscation: but strongly close the mouth of the vessel, and cook for forty days, when ye will find it water whiter than milk; deprive it of all blackness by cooking; continue the cooking until its whole nature be disintegrated, until the defilement perish, until it be found clean, and is wholly broken up (or becomes wholly clean). But if ye wish that the whole arcanum, which I have given you, be accomplished, wash the same with water, that is to say, the other part which I counselled you to preserve, until there appear a crocus, and leave in its own vessel. For the Iksir pounds (or contains) itself; imbue also with the residue of the water, until by decoction and by water it be pounded and become like a syrup of pomegranates; imbue it, therefore, and cook, until the weight of the humidity shall fail, and the colour which the Philosophers have magnified shall truly appear.

The Twenty-third Dictum.

Cerus saith: Understand, all ye Sons of the Doctrine, that which Theophilus hath told you, namely, that there exists an affinity between the magnet and the iron, by the alliance of composite existing between the magnet and the iron, while the copper is fitly ruled for one hundred days: what statement can be more useful to you than that there is no affinity between tin and quicksilver!

The Turba answereth: Thou hast ill spoken, having disparaged the true disposition.

And he: I testify that I say nothing but what is true why are you incensed against me Fear the Lord, all ye Turba, that you Master may believe you!

The Turba answereth: Say what you will.

And he: I direct you to take quicksilver, in which is the male potency or strength; cook the same with its body until it becomes a fluxible water; cook the masculine together with the vapour, until each shall be coagulated and become a stone. Then take the water which you had divided into two parts, of which one is for liquefying and cooking the body, but the second is for cleansing that which is already burnt, and its companion, which [two] are made one. Imbue the stone seven times, and cleanse, until it be disintegrated, and its body be purged from all defilement, and become earth. Know also that in the time of forty-two days the whole is changed into earth; by cooking, therefore, liquefy the same until it become as true water, which is quicksilver. Then wash with water of nitre until it become as a liquefied coin. Then cook until it be congealed and become like to tin, when it is a most great arcanum; that is to say, the stone which is out of two things. Rule the same by cooking and pounding, until it becomes a most excellent crocus. Know also that unto water desiccated with its companion we have given the name of crocus. Cook it, therefore, and imbue with the residual water reserved by you until you attain your purpose.

The Twenty-fourth Dictum.

Bocascus saith: Thou hast spoken well, O Belus, and therefore I follow thy steps!

He answereth: As it may please you, but do not become envious, for that is not the part of the Wise.

And Bocascus: Thou speakest the truth, and thus, therefore, I direct the Sons of the Doctrine. Take lead, and, as the Philosophers have ordained, imbue, liquefy, and afterwards congeal, until a stone is produced; then rule the stone with gluten of gold and syrup of pomegranates until it be broken up. But you have already divided the water into two parts, with one of which you have liquefied the lead, and it has become as water; cook, therefore, the same until it be dried and have become earth; then pound with the water reserved until it acquire a red colour, as you have been frequently ordered.

The Turba answereth: Thou hast done nothing but pile up ambiguous words. Return, therefore, to the subject.

And he: Ye who wish to coagulate quicksilver, must
mix it with its equal. Afterwards cook it diligently
until both become permanent water, and, again, cook
this water until it be coagulated. But let this be
desiccated with its own equal vapour, because ye
have found the whole quicksilver to be coagulated by
itself. If ye understand, and place in your vessel
what is necessary, cook it until it be coagulated,
and then pound until it becomes a crocus like to the
colour of gold.

The Twenty-fifth Dictum.

Menabdus saith: May God reward thee for the regimen, since thou speakest the truth! For thou hast illuminated thy words.

And they: It is said because thou praisest him for his sayings, do not be inferior to him.

And he: I know that I can utter nothing but that which he hath uttered; however, I counsel posterity to make bodies not bodies, but these incorporeal things bodies. For by this regimen the composite is prepared, and the hidden part of its nature is extracted. With these bodies accordingly join quicksilver and the body of Magnesia, the woman also with the man, and by means of this there is extracted our secret Ethelia, through which bodies are coloured; assuredly, if I understand this regimen, bodies become not bodies, and incorporeal things become bodies. If ye diligently pound the things in the fire and digest (or join to) the Ethelias, they become clean and fixed things. And know ye that quicksilver is a fire burning the bodies, mortifying and breaking up, with one regimen, and the more it is mixed and pounded with the body, the more the body is disintegrated, while the quicksilver is attenuated and becomes living. For when ye shall diligently pound fiery quicksilver and cook it as required, ye will possess Ethel, a fixed nature and colour, subject to every tincture, which also overcomes, breaks, and constrains the fire. For this reason it does not colour things unless it be coloured, and being coloured it colours. And know that no body can tinge itself unless its spirit be extracted from the secret belly thereof, when it becomes a body and soul without the spirit, which is a spiritual tincture, out of which colours have manifested, seeing that a dense thing does not tinge a tenuous, but a tenuous nature colours that which enters into a body. When, however, ye have ruled the body of copper, and have extracted from it a most tenuous (subject), then the latter is changed into a tincture by which it is coloured. Hence has the wise man said, that copper does not tinge unless first it be tinged. And know that those four bodies which you are directed to rule are this copper, and that the tinctures which I have signified unto you are the condensed and the humid, but the condensed is a conjoined vapour, and the humid is the water of sulphur, for sulphurs are contained by sulphurs, and rightly by these things Nature rejoices in Nature, and overcomes, and constrains.

The Twenty-Sixth Dictum.

Zenon saith: I perceive that you, O crowd of the Wise, have conjoined two bodies, which your Master by no means ordered you to do!

The Turba answereth: Inform us according to your own opinion, O Zenon, in this matter, and beware of envy! Then he: Know that the colours which shall appear to you out of it are these. Know, O Sons of the Doctrine, that it behoves you to allow the composition to putrefy for forty days, and then to sublimate five times in a vessel. Next join to a fire of dung, and cook, when these colours shall appear to you: On the first day black citrine, on the second black red, on the third like unto a dry crocus, finally, the purple colour will appear to you; the ferment and the coin of the vulgar shall be imposed; then is the Ixir composed out of the humid and the dry, and then it tinges with an invariable tincture. Know also that it is called a body wherein there is gold. But when ye are composing the Ixir, beware lest you extract the same hastily, for it lingers. Extract, therefore, the same as an Ixir. For this venom is, as it were, birth and life, because it is a soul extracted out of many things, and imposed upon coins: its tincture, therefore, is life to those things with which it is joined, from

which it removes evil, but it is death to the bodies
from which it is extracted. Accordingly, the Masters
have said that between them there exists the same
desire as between male and female, and if any one,
being introduced to this Art, should know these
natures, he would sustain the tediousness of cooking
until he gained his purpose according to the will of
God.

The Twenty-Seventh Dictum.

Gregorius saith: O all ye Turba, it is to be observed that the envious have called the venerable stone Efflucidinus, and they have ordered it to be ruled until it coruscates like marble in its splendour.

And they: Show, therefore, what it is to posterity.

Then he: Willingly; you must know that the copper is commingled with vinegar, and ruled until it becomes water. Finally, let it be congealed, and it remains a coruscating stone with a brilliancy like marble, which, when ye see thus, I direct you to rule until it becomes red, because when it is cooked till it is disintegrated and becomes earth, it is turned into a red colour. When ye see it thus, repeatedly cook and imbue it until it assume the aforesaid colour, and it shall become hidden gold. Then repeat the process, when it will become gold of a Tyrian colour. It behoves you, therefore, O all ye investigators of this Art, when ye have observed that this Stone is coruscating, to pound and turn it into earth, until it acquires some degree of redness; then take the remainder of the water which the envious ordered you to divide into two parts, and ye shall imbibe them several times until the colours which are hidden by no body appear unto you. Know also that if ye rule it ignorantly, ye shall see nothing of those colours. I knew a certain person who commenced this work, and operated the natures of truth, who, when the redness was somewhat slow in appearing, imagined that he had made a mistake, and so relinquished the work. Observe, therefore, how ye make the conjunction, for the punic dye, having embraced his spouse, passes swiftly into her body, liquefies, congeals, breaks up, and disintegrates the same. Finally, the redness does not delay in coming, and if ye effect it without the weight, death will take place, whereupon it will be thought to be bad. Hence, I order that the fire should be gentle in liquefaction, but when it is turned to earth make the same intense, and imbue it until God shall extract the colours for us and they appear.

The Twenty-Eighth Dictum.

Custos saith: I am surprised, O all ye Turba! at the very great force and nature of this water, for when it has entered into the said body, it turns it first into earth, and next into powder, to test the perfection of which take in the hand, and if ye find it impalpable as water, it is then most excellent; otherwise, repeat the cooking until it is brought to the required condition. And know that if ye use any substance other than our copper, and rule with our water, it will profit you nothing. If, on the other hand, ye rule our copper with our water, ye shall find all that has been promised by us.

But the Turba answereth: Father, the envious created no little obscurity when they commanded us to take lead and white quicksilver, and to rule the same with dew and the sun till it becomes a coin-like stone.

Then he: They meant our copper and our permanent water, when they thus directed you to cook in a gentle fire, and affirmed that there should be produced the said coin-like stone, concerning which the Wise have also observed, that Nature rejoices in Nature, by reason of the affinity which they know to exist between the two bodies, that is to say, copper and permanent water. Therefore, the nature of these two is one, for between them there is a mixed affinity, without which they would not so swiftly unite, and be held together so that they may become one.

Saith the Turba: Why do the envious direct us to take the copper which we have now made, and roasted until it has become gold!

The Twenty-Ninth Dictum.

Diamedes saith: Thou hast spoken already, O Moses [Custos], in an ungrudging manner, as became thee; I will also confirm thy words, passing over the hardness of the elements which the wise desire to remove, this disposition being most precious in their eyes. Know, O ye seekers after this doctrine, that man does not proceed except from a man; that only which is like unto themselves is begotten from brute animals; and so also with flying creatures.

I have treated these matters in compendious fashion, exalting you towards the truth, who yourselves omit prolixity, for Nature is truly not improved by Nature, save with her own nature, seeing that thou thyself art not improved except in thy son, that is to say, man in man. See, therefore, that ye do not neglect the precepts concerning her, but make use of venerable Nature, for out of her Art cometh, and out of no other. Know also that unless you seize hold of this Nature and rule it, ye will obtain nothing. Join, therefore, that male, who is son to the red slave, in marriage with his fragrant wife, which having been done, Art is produced between them; add no foreign matter unto these things, neither powder nor anything else; that conception is sufficient for us, for it is near, yet the son is nearer still. How exceeding precious is the nature of that red slave, without which the regimen cannot endure!

Bacsen saith: O Diomedes, thou hast publicly revealed this disposition!

He answereth: I will even shed more light upon it. Woe unto you who fear not God, for He may deprive you of this art! Why, therefore, are you envious towards your brethren?

They answer: We do not flee except from fools; tell us, therefore, what is thy will?

And he: Place Citrine with his wife after the conjunction into the bath; do not kindle the bath excessively, lest they be deprived of sense and motion; cause them to remain in the bath until their body, and the colour thereof, shall become a certain unity, whereupon restore unto it the sweat thereof; again suffer it to die; then give it rest, and beware lest ye evaporate them by burning them in too strong a fire. Venerate the king and his wife, and do not burn them, since you know not when you may have need of these things, which improve the king and his wife. Cook them, therefore, until they become black, then white, afterwards red, and finally until a tingeing venom is produced. O seekers after this Science, happy are ye, if ye understand, but if not, I have still performed my duty, and that briefly, so that if ye, remain ignorant, it is God who hath concealed the truth from you! Blame not, therefore, the Wise, but yourselves, for if God knew that ye possessed a faithful mind, most certainly he would reveal unto you the truth. Behold, I have established you therein, and have extricated you from error!

The Thirtieth Dictum.

Bacsen saith: Thou hast spoken well, O Diomedes, but I do not see that thou hast demonstrated the disposition of Corsufle to posterity! Of this same Corsufle the envious have spoken in many ways, and have confused it with all manner of names.

Then he: Tell me, therefore, O Bacsen, according to thy opinion in these matters, and I swear by thy father that this is the head of the work, for the true beginning hereof cometh after the completion.

Bacsen saith: I give notice, therefore, to future seekers after this Art, that Corsufle is a composite, and that it must be roasted seven times, because when it arrives at perfection it tinges the whole body.

The Turba answereth: Thou hast spoken the truth, O Bacsen!

The Thirty-First Dictum.

Pythagoras Saith: How does the discourse of Bacsen appear to you, since he has omitted to name the substance by its artificial names?

And they: Name it, therefore, oh Pythagoras!

And he: Corsufle being its composition, they have applied to it all the names of bodies in the world, as, for example, those of coin, copper, tin, gold, iron, and also the name of lead, until it be deprived of that colour and become Ixir.

The Turba answereth: Thou hast spoken well, O Pythagoras!

And he: Ye have also spoken well, and some among the others may discourse concerning the residual matters.

The Thirty-Second Dictum.

Bonellus saith: According to thee, O Pythagoras, all things die and live by the will of God, because that nature from which the humidity is removed, that nature which is left by nights, does indeed seem like unto something that is dead; it is then turned and (again) left for certain nights, as a man is left in his tomb, when it becomes a powder. These things being done, God will restore unto it both the soul and the spirit thereof, and the weakness being taken away, that matter will be made strong, and after corruption will be improved, even as a man becomes stronger after resurrection and younger than he was in this world. Therefore it behoves you, O ye Sons of the Doctrine, to consume that matter with fire boldly until it shall become a cinder, when know that ye have mixed it excellently well, for that cinder receives the spirit, and is imbued with the humour until it assumes a fairer colour than it previously possessed. Consider, therefore, O ye Sons of the Doctrine, that artists are unable to paint with their own tinctures until they convert them into a powder; similarly, the philosophers cannot combine medicines for the sick slaves until they also turn them into powder, cooking some of them to a cinder, while others they grind with their hands.

The case is the same with those who compose the
images of the ancients. But if ye understand what
has already been said, ye will know that I speak the
truth, and hence I have ordered you to burn up the
body and turn it into a cinder, for if ye rule it
subtly many things will proceed from it, even as
much proceeds from the smallest things in the world.
It is thus because copper like man, has a body and a
soul, for the inspiration of men cometh from the
air, which after God is their life, and similarly
the copper is inspired by the humour from which that
same copper receiving strength is multiplied and
augmented like other things. Hence, the philosophers
add, that when copper is consumed with fire and
iterated several times, it becomes better than it
was.

The Turba answereth: Show, therefore, O Bonellus, to
future generations after what manner it becometh
better than it was!

And he: I will do so willingly; it is because it is
augmented and multiplied, and because God extracts
many things out of one thing, since He hath created
nothing which wants its own regimen, and those
qualities by which its healing must be effected.
Similarly, our copper, when it is first cooked,
becomes water; then the more it is cooked, the more
is it thickened until it becomes a stone, as the

envious have termed it, but it is really an egg tending to become a metal. It is afterwards broken and imbued, when ye must roast it in a fire more intense than the former, until it shall be coloured and shall become like blood in combustion, when it is placed on coins and changes them into gold, according to the Divine pleasure. Do you not see that sperm is not produced from the blood unless it be diligently cooked in the liver till it has acquired an intense red colour, after which no change takes place in that sperm? It is the same with our work, for unless it be cooked diligently until it shall become a powder, and afterwards be putrefied until it shall become a spiritual sperm, there will in no wise proceed from it that colour which ye desire. But if ye arrive at the conclusion of this regimen, and so obtain your purpose, ye shall be princes among the People of your time.

The Thirty-Third Dictum.

Nicarus saith: Now ye have made this arcanum public.

The Turba answereth: Thus did the Master order.

And he: Not the whole, nevertheless.

But they: He ordered us to clear away the darkness therefrom; do thou, therefore, tell us.

And he: I counsel posterity to take the gold which they wish to multiply and renovate, then to divide the water into two parts.

And they: Distinguish, therefore, when they divide the water.

But he: It behoves them to burn up our copper with one part. For the said copper, dissolved in that water, is called the ferment of Gold, if ye rule well. For the same in like manner are cooked and liquefy as water; finally, by cooking they are congealed, crumble, and the red appears. But then it behoves you to imbue seven times with the residual water, until they absorb all the water, and, all the moisture being dried up, they are turned into dry earth; then kindle a fire and place therein for forty days until the whole shall putrefy, and its colours appear.

The Thirty-Fourth Dictum.

Bacsen saith: On account of thy dicta the Philosophers said beware. Take the regal Corsufle, which is like to the redness of copper, and pound in the urine of a calf until the nature of the Corsufle is converted, for the true nature has been hidden in the belly of the Corsufle.

The Turba saith: Explain to posterity what the nature is.

And he: A tingeing spirit which it hath from permanent water, which is coin-like, and coruscates.

And they: Shew, therefore, how it is extracted.

And he: It is pounded, and water is poured upon it seven times until it absorbs the whole humour, and receives a force which is equal to the hostility of the fire; then it is called rust. Putrefy the same diligently until it becomes a spiritual powder, of a colour like burnt blood, which the fire overcoming hath introduced into the receptive belly of Nature, and hath coloured with an indelible colour. This, therefore, have kings sought, but not found, save only to whom God has granted it.

But the Turba saith: Finish your speech, O Bacsen.

And he: I direct them to whiten copper with white water, by which also they make red. Be careful not to introduce any foreign matter.

And the Turba: Well hast thou spoken, O Bacsen, and Nictimerus also has spoken well!

Then he: If I have spoken well, do one of you continue.

The Thirty-Fifth Dictum.

But Zimon saith: Hast thou left anything to be said by another?

And the Turba: Since the words of Nicarus and Bacsen are of little good to those who seek after this Art, tell us, therefore, what thou knowest, according as we have said.

And he: Ye speak the truth, O all ye seekers after this Art! Nothing else has led you into error but the sayings of the envious, because what ye seek is sold at the smallest possible price. If men knew this, and how great was the thing they held in their hands, they would in no wise sell it. Therefore, the Philosophers have glorified that venom, have treated of it variously, and in many ways, have taken and applied to it all manner of names, wherefore, certain envious persons have said: It is a stone and not a stone, but a gum of Ascotia, consequently, the Philosophers have concealed the power thereof. For this spirit which ye seek, that ye may tinge therewith, is concealed in the body, and hidden away from sight, even as the soul in the human body. But ye seekers after the Art, unless ye disintegrate this body, imbue and pound both cautiously and diligently, until ye extract it from its grossness (or grease), and turn it into a tenuous and impalpable spirit, have your labour in vain. Wherefore the Philosophers have said: Except ye turn bodies into not bodies, and incorporeal things into bodies, ye have not yet discovered the rule of operation.

But the Turba saith: Tell, therefore, posterity how bodies are turned into not-bodies.

And he: They are pounded with fire and Ethelia till they become a powder. And know that this does not take place except by an exceedingly strong decoction, and continuous contrition, performed with a moderate fire, not with hands, with imbibition and putrefaction, with exposure to the sun and to Ethelia. The envious caused the vulgar to err in this Art when they stated that the thing is common in its nature and is sold at a small price. They further said that the nature was more precious than all natures, wherefore they deceived those who had recourse to their books. At the same time they spoke the truth, and therefore doubt not these things.

But the Turba answereth: Seeing that thou believest the sayings of the envious, explain, therefore, to posterity the disposition of the two natures.

And he: I testify to you that Art requires two natures, for the precious is not produced without the common, nor the common without the precious. It behoves you, therefore, O all ye Investigators of this Art, to follow the sayings of Victimerus, when he said to his disciples: Nothing else helps you save to sublimate water and vapour.

And the Turba: The whole work is in the vapour and the sublimation of water. Demonstrate, therefore, to them the disposition of the vapour.

And he: When ye shall perceive that the natures have become water by reason of the heat of the fire, and that they have been purified, and that the whole body of Magnesia is liquefied as water; then all things have been made vapour, and rightly, for then the vapour contains its own equal, wherefore the envious call either vapour, because both are joined in decoctions, and one contains the other. Thus our stag finds no path to escape, although flight be essential to it. The one keeps back the other, so that it has no opportunity to fly, and it finds no place to escape; hence all are made permanent, for when the one falls, being hidden in the body, it is congealed with it, and its colour varies, and it extracts its nature from the properties which God has infused into His elect, and it alienates it, lest it flee. But the blackness and redness appear,

94

and it falls into sickness, and dies by rust and putrefaction; properly speaking, then, it has not a flight, although it is desirous to escape servitude; then when it is free it follows its spouse, that a favourable colour may befall itself and its spouse; its beauty is not as it was, but when it is placed with coins, it makes them gold. For this reason, therefore, the Philosophers have called the spirit and the soul vapour. They have also called it the black humid wanting perlution; and forasmuch as in man there are both humidity and dryness, thus our work, which the envious have concealed, is nothing else but vapour and water.

The Turba answereth: Demonstrate vapour and water!

And he: I say that the work is out of two; the envious have called it composed out of two, because these two become four, wherein are dryness and humidity, spirit and vapour.

The Turba answereth: Thou hast spoken excellently, and without envy. Let Zimon next follow.

The Thirty-Sixth Dictum.

Afflontus, the Philosopher, saith: I notify to you all, O ye investigators of this Art, that unless ye sublime the substances at the commencement by cooking, without contrition of hands, until the whole become water, ye have not yet found the work. And know ye, that the copper was formerly called sand, but by others stone, and, indeed, the names vary in every regimen. Know further, that the nature and humidity become water, then a stone, if ye cause them to be well complexionated, and if ye are acquainted with the natures, because the part which is light and spiritual rises to the top, but that which is thick and heavy remains below in the vessel. Now this is the contrition of the Philosophers, namely, that which is not sublimated sinks down, but that which becomes a spiritual powder rises to the top of the vessel, and this is the contrition of decoction, not of hands. Know also, that unless ye have turned all into powder, ye have not yet pounded them completely. Cook them, therefore, successively until they become converted, and a powder. Wherefore Agadaimon saith: Cook the copper until it become a gentle and impalpable body, and impose in its own vessel; then sublimate the same six or seven times until the water shall descend. And know that when the water has become powder then has it been ground diligently. But if ye ask, how is the water made a powder? note that the intention of the Philosophers is that the body before which before it falls into the water is not water may become water; the said water is mixed with the other water, and they become one water. It is to be stated, therefore, that unless ye turn the thing mentioned into water, ye shall not attain to the work. It is, therefore, necessary for the body to be so possessed by the flame of the fire that it is disintegrated and becomes weak with the water, when the water has been added to the water, until the whole becomes water. But fools, hearing of water, think that this is water of the clouds. Had they read our books they would know that it is permanent water, which cannot become permanent without its companion, wherewith it is made one. But this is the water which the Philosophers have called Water of Gold, the Igneous, Good Venom, and that Sand of Many Names which Hermes ordered to be washed frequently, so that the blackness of the Sun might be removed, which he introduced in the solution of the body. And know, all ye seekers after this Art, that unless ye take this pure body, that is, our copper without the spirit, ye will by no means see what ye desire, because no foreign thing enters therein, nor does anything enter unless it be pure. Therefore, all ye seekers after this Art, dismiss the multitude of obscure names, for the nature is one water; if anyone err, he draws nigh to destruction, and loses his life. Therefore, keep this one nature, but dismiss what is foreign.

The Thirty-Seventh Dictum.

Bonellus saith: I will speak a little concerning Magnesia.

The Turba answereth: Speak.

And he: O all ye Sons of the Doctrine, when mixing Magnesia, place it in its vessel, the mouth of which close carefully, and cook with a gentle fire until it liquefy, and all become water therein! For the heat of the water acting thereupon, it becomes water by the will of God. When ye see that the said water is about to become black, ye know that the body is already liquefied. Place again in its vessel, and cook for forty days, until it drink up the moisture of the vinegar and honey. But certain persons uncover it, say, once in each week, or once in every ten nights; in either case, the ultimate perfection of pure water appears at the end of forty days, for then it completely absorbs the humour of the decoction. Therefore, wash the same, and deprive of its blackness, until, the blackness being removed, the stone becomes dry to the touch. Hence the envious have said: Wash the Magnesia with soft water, and cook diligently, until it become earth, and the humour perish. Then it is called copper. Subsequently, pour very sharp vinegar upon it, and leave it to be soaked therein. But this is our copper, which the Philosophers have ordained should be washed with permanent water, wherefore they have said: Let the venom be divided into two parts, with one of which burn up the body, and with the other putrefy. And know, all ye seekers after this Science, that the whole work and regimen does not take place except by water, wherefore, they say that the thing which ye seek is one, and, unless that which improves it be present in the said thing, what ye look for shall in no wise take place. Therefore, it behoves you to add those .things which are needful, that ye may thereby obtain that which you purpose.

The Turba answereth: Thou has spoken excellently, O Bonellus! If it please thee, therefore, finish that which thou art saying; otherwise repeat it a second time.

But he: Shall I indeed repeat these and like things? O all ye investigators of this Art, take our copper; place with the first part of the water in the vessel; cook for forty days; purify from all uncleanliness; cook further until its days be accomplished, and it become a stone having no moisture. Then cook until nothing remains except faeces. This done, cleanse seven times, wash with water, and when the water is used up leave it to putrefy in its vessel, so long as may seem desirable to your purpose. But the envious called this composition when it is turned into blackness that which is sufficiently black, and have said: Rule the same with vinegar and nitre. But that which remained when it had been whitened they called sufficiently white, and ordained that it should be ruled with permanent water. Again, when they called the same sufficiently red, they ordained that it should be ruled with water and fire until it became red.

The Turba answereth: Show forth unto posterity what they intended by these things.

And he: They called it Ixir satis, by reason of the variation of its colours. In the work, however, there is neither variety, multiplicity, nor opposition of substances; it is necessary only to make the black copper white and then red. However, the truth-speaking Philosophers had no other intention than that of liquefying, pounding, and cooking Ixir until the stone should become like unto marble in its splendour. Accordingly, the envious again said: Cook the same with vapour until the stone becomes coruscating by reason of its brilliancy. But when ye see it thus, it is, indeed, the most great Arcanum. Notwithstanding, ye must then pound and wash it seven times with permanent water; finally, again pound and congeal in its own water, until ye extract its own concealed nature. Wherefore, saith Maria, sulphurs are contained in sulphurs, but humour in like humour, and out of sulphur mixed with sulphur, there comes forth a great work. But I ordain that you rule the same with dew and the sun, until your purpose appear to you. For I signify unto you that there are two kinds of whitening and of making red, of which one consists in rust and the other in contrition and decoction. But ye do not need any contrition of hands. Beware, however, of making a separation from the waters lest the poisons get at You, and the body perish with the other things which are in the vessel.

The Thirty-Eighth Dictum.

Effistus saith: Thou hast spoken most excellently, O Bonellus, and I bear witness to all thy words!

The Turba saith: Tell us if there be any service in the speech of Bonellus, so that those initiated in this disposition may be more bold and certain.

Effistus saith: Consider, all ye investigators of this Art, how Hermes, chief of the Philosophers, spoke and demonstrated when he wished to mix the natures. Take, he tells us, the stone of gold, combine with humour which is permanent water, set in its vessel, over a gentle fire until liquefaction takes place. Then leave it until the water dries, and the sand and water are combined, one with another; then let the fire be more intense than before, until it again becomes dry, and is made earth. When this is done, understand that here is the beginning of the arcanum; but do this many times, until two-thirds of the water perish, and colours manifest unto you.

The Turba answereth: Thou hast spoken excellently, O Effistus! Yet, briefly inform us further.

And he: I testify to Posterity that the dealbation doth not take place save by decoction. Consequently,

Agadaimon has very properly treated of cooking, of pounding, and of imbuing, ethelia. Yet I direct you not to pour on the whole of the water at one time, lest the Ixir be submerged, but pour it in gradually, pound and dessicate, and do this several times until the water be exhausted. Now concerning this the envious have said: Leave the water when it has all been poured in, and it will sink to the bottom. But their intention is this, that while the humour is drying, and when it has been turned into powder, leave it in its glass vessel for forty days, until it passes through various colours, which the Philosophers have described. By this method of cooking the bodies put on their spirits and spiritual tinctures, and become warm.

The Turba answereth: Thou hast given light to us, O Effistus, and hast done excellently! Truly art thou cleared from envy; wherefore, let one of you others speak as he pleases.

The Thirty-Ninth Dictum.

Bacsen saith: O all ye seekers after this Art, ye can reach no useful result without a patient, laborious, and solicitous soul, persevering courage, and continuous regimen. He, therefore, who is willing to Persevere in this disposition, and would enjoy the result, may enter upon it, but he who desires to learn over speedily, must not have recourse to our books, for they impose great labour before they are read in their higher sense, once, twice, or thrice. Therefore, the Master saith: Whosoever bends his back over the study of our books, devoting his leisure thereto, is not occupied with vain thoughts, but fears God, and shall reign in the Kingdom without fail until he die. For what ye seek is not of small price. Woe unto you who seek the very great and compensating treasure of God! Know ye not that for the smallest Purpose in the world, earthly men will give themselves to death, and what, therefore, ought they to do for this most excellent and almost impossible offering? Now, the regimen is greater than is perceived by reason, except through divine inspiration. I once met with a person who was as well acquainted with the elements as I myself, but when he proceeded to rule this disposition, he attained not to the joy thereof by reason of his sadness and ignorance in ruling, and excessive eagerness, desire, and haste concerning the purpose. Woe unto you, sons of the Doctrine! For one who plants trees does not look for fruit, save in due season; he also who sows seeds does not expect to reap, except at harvest time. How, then, should ye desire to attain this offering when ye have read but a single book, or have adventured only the first regimen? But the Philosophers have plainly stated that the truth is not to be discerned except after error, and nothing creates greater pain at heart than error in this Art, while each imagines that he has almost the whole world, and yet finds nothing in his hands. Woe unto you! Understand the dictum of the Philosopher, and how he divided the work when he said- pound, cook, reiterate, and be thou not weary. But when thus he divided the work, he signified commingling, cooking, assimilating, roasting, heating, whitening, pounding, cooking Ethelia, making rust or redness, and tingeing. Here, therefore, are there many names, and yet there is one regimen. And if men knew that one decoction and one contrition would suffice them, they would not so often repeat their words, as they have done, and in order that the mixed body may be pounded and cooked diligently, have admonished you not to be weary thereof. Having darkened the matter to you with their words, it suffices me to speak in this manner.

It is needful to complexionate the venom rightly, then cook many times, and do not grow tired of the decoction. Imbue and cook it until it shall become as I have ordained that it should be ruled by you-namely, impalpable spirits, and until ye perceive that the Ixir is clad in the garment of the Kingdom. For when ye behold the Ixir turned into Tyrian colour, then have ye found that which the Philosophers discovered before you. If ye understand my words (and although my words be dead, yet is there life therein for those who understand themselves), they will forthwith explain any ambiguity occurring herein. Read, therefore, repeatedly, for reading is a dead speech, but that which is uttered with the lips the same is living speech. Hence we have ordered you to read frequently, and, moreover, ponder diligently over the things which we have narrated.

The Fortieth Dictum.

Jargus saith: Thou hast left obscure a part of thy discourse, O Bacsen!

And he: Do thou, therefore, Jargus, in thy clemency shew forth the same!

And he answereth: The copper of which thou hast before spoken is not copper, nor is it the tin of the vulgar; it is our true work (or body) which must be combined with the body of Magnesia, that it may be cooked and pounded without wearying until the stone is made. Afterwards, that stone must be pounded in its vessel with the water of nitre, and, subsequently, placed in liquefaction until it is destroyed. But, all ye investigators of this art, it is necessary to have a water by which the more you cook, so much the more you sprinkle, until the said copper shall put on rust, which is the foundation of our work. Cook, therefore, and pound with Egyptian vinegar.

The Forty-First Dictum.

Zimon saith: Whatsoever thou hast uttered, O Jargos, is true, yet I do not see that the whole Turba hath spoken concerning the rotundum.

Then he: Speak, therefore, thine opinion concerning it, O Zimon!

Zimon saith: I notify to Posterity that the rotundum turns into four elements, and is derived out of one thing.

The Turba answereth: Inasmuch as thou art speaking, explain for future generations the method of ruling.

And he: Willingly: it is necessary to take one part of our copper, but of Permanent Water three parts; then let them be mixed and cooked until they be thickened and become one stone, concerning which the envious have said: Take one part of the pure body, but three parts of copper of Magnesia; then commingle with rectified vinegar, mixed with male of earth; close the vessel, observe what is in it, and cook continuously until it becomes earth.

The Forty-Second Dictum.

Ascanius saith: Too much talking, O all ye Sons of the Doctrine, leads this subject further into error! But when ye read in the books of the Philosophers that Nature is one only, and that she overcomes all things: Know that they are one thing and one composite. Do ye not see that the complexion of a man is formed out of a soul and body; thus, also, must ye conjoin these, because the Philosophers, when they prepared the matters and conjoined spouses mutually in love with each other, behold there ascended from them a golden water!

The Turba answereth: When thou wast treating of the first work, lo! thou didst turn unto the second! How ambiguous hast thou made thy book, and how obscure are thy words!

Then he: I will perform the disposition of the first work.

The Turba answereth: Do this.

And he: Stir up war between copper and quicksilver, until they go to destruction and are corrupted, because when the copper conceives the quicksilver it coagulates it, but when the quicksilver conceives the copper, the copper is congealed into earth; stir up, therefore, a fight between them; destroy the body of the copper until it becomes a powder. But conjoin the male to the female, which are vapour and quicksilver, until the male and the female become Ethel, for he who changes them into spirit by means of Ethel, and next makes them red, tinges every body, because, when by diligent cooking ye pound the body, ye extract a pure, spiritual, and sublime soul therefrom, which tinges every body.

The Turba answereth: Inform, therefore, posterity what is that body.

And he: It is a natural sulphureous thing which is called by the names of all bodies.

The Forty-Third Dictum.

Dardaris saith: Ye have frequently treated of the regimen, and have introduced the conjunction, yet I proclaim to posterity that they cannot extract the now hidden soul except by Ethelia, by which bodies become not bodies through continual cooking, and by sublimation of Ethelia. Know also that quicksilver is fiery, burning every body more than does fire, also mortifying bodies, and that every body which is mingled with it is ground and delivered over to be destroyed. When, therefore, ye have diligently pounded the bodies, and have exalted them as required, therefrom is produced that Ethel nature, and a colour which is tingeing and not volatile, and it tinges the copper which the Turba said did not tinge until it is tinged, because that which is tinged tinges. Know also that the body of the copper is ruled by Magnesia, and that quicksilver is four bodies, also that the matter has no being except by humidity, because it is the water of sulphur, for sulphurs are contained in sulphurs.

The Turba saith: O Dardaris, inform posterity what sulphurs are!

And he: Sulphurs are souls which are hidden in four bodies, and, extracted by themselves, do contain one another, and are naturally conjoined. For if ye rule that which is hidden in the belly of sulphur with water, and cleanse well that which is hidden, then nature rejoices, meeting with nature, and water similarly with its equal. Know ye also that the four bodies are not tinged but tinge.

And the Turba: Why dost thou not say like the ancients that when they are tinged, they tinge?

And he: I state that the four coins of the vulgar populace are not tinged, but they tinge copper, and when that copper is tinged, it tinges the coins of the populace.

The Forty-Fourth Dictum.

Moyses saith: This one thing of which thou hast told us, O Dardaris, the Philosophers have called by many names, sometimes by two and sometimes by three names!

Dardaris answereth: Name it, therefore, for posterity, setting aside envy.

And he: The one is that which is fiery, the two is the

body composed in it, the three is the water of sulphur, with which also it is washed and ruled until it be perfected. Do ye not see what the Philosopher affirms, that the quicksilver which tinges gold is quicksilver out of Cambar?

Dardaris answereth: What dost thou mean by this? For the Philosopher says: sometimes from Cambar and sometimes from Orpiment.

And he: Quicksilver of orpiment is Cambar of Magnesia, but quicksilver is sulphur ascending from the mixed composite. Ye must, therefore, mix that thick thing with fiery venom, putrefy, and diligently pound until a spirit be produced, which

is hidden in that other spirit; then is made the
tincture which is desired of you all.

The Forty-Fifth Dictum.

But Plato saith: It behoves you all, O Masters, when those bodies are being dissolved, to take care lest they be burnt up, as also to wash them with sea water, until all their salt be turned into sweetness, clarifies, tinges, becomes tincture of copper, and then goes off in flight! Because it was necessary that one should become tingeing, and that the other should be tinged, for the spirit being separated from the body and hidden in the other spirit, both become volatile. Therefore the Wise have said that the gate of flight must not be opened for that which would flee, (or that which does not flee), by whose flight death is occasioned, for by the conversion of the sulphureous thing into a spirit like unto itself, either becomes volatile, since they are made aeriform spirits prone to ascend in the air. But the Philosophers seeing that which was not volatile made volatile with the volatiles, iterated these to a body like to the non-volatiles, and put them into that from which they could not escape. They iterated them to a body like unto the bodies from which they were extracted, and the same were then digested. But as for the statement of the Philosopher that the tingeing agent and that which is to be tinged are made one tincture, it refers to a spirit concealed in another humid spirit. Know also that one of the humid spirits is cold, but the other is hot, and although the cold humid is not adapted to the warm humid, nevertheless they are made one. Therefore, we prefer these two bodies, because by them we rule the whole work, namely, bodies by not-bodies, until incorporeals become bodies, steadfast in the fire, because they are conjoined with volatiles, which is not possible in any body, these excepted. For spirits in every wise avoid bodies, but fugitives are restrained by incorporeals. Incorporeals, therefore, similarly flee from bodies; those, consequently, which do not flee are better and more precious than all bodies. These things, therefore, being done, take those which are not volatile and join them; wash the body with the incorporeal until the incorporeal receives a non-volatile body; convert the earth into water, water into fire, fire into air, and conceal the fire in the depths of the water, but the earth in the belly of the air, mingling the hot with the humid, and the cold with the dry. Know, also, that Nature overcomes Nature, Nature rejoices in Nature, Nature contains Nature.

The Forty-Sixth Dictum.

Attamus saith: It is to be noted that the whole assembly of the Philosophers have frequently treated concerning

Rubigo. Rubigo, however, is a fictitious and not a true name.

The Turba answereth: Name, therefore, Rubigo by its true name, for by this it is not calumniated.

And he: Rubigo is according to the work, because it is from gold alone.

The Turba answereth: Why, then, have the Philosophers referred it to the leech?

He answereth: Because water is hidden in sulphureous gold as the leech is in water; rubigo, therefore, is rubefaction in the second work, but to make rubigo is to whiten in the former work, in which the Philosophers ordained that the flower of gold should be taken and a proportion of gold equally.

The Forty-Seventh Dictum.

Mundus saith: Thou hast already treated sufficiently of Rubigo, O Attamus! I will speak, therefore, of venom, and will instruct future generations that venom is not a body, because subtle spirits have made it into a tenuous spirit, have tinged the body and burned it with venom, which venom the Philosopher asserts will tinge every body. But the Ancient Philosophers thought that he who turned gold into venom had arrived at the purpose, but he who can do not this profiteth nothing. Now I say unto you, all ye Sons of the Doctrine, that unless ye reduce the thing by fire until those things ascend like a spirit, ye effect nought. This, therefore, is a spirit avoiding the fire and a ponderous smoke, which when it enters the body penetrates it entirely, and makes the body rejoice. The Philosophers have all said: Take a black and conjoining spirit; therewith break up the bodies and torture them till they be altered.

The Forty-Eighth Dictum.

Pythagoras saith: We must affirm unto all you seekers after this Art that the Philosophers have treated of conjunction (or continuation) in various ways. But I enjoin upon you to make quicksilver con strain the body of Magnesia, or the body Kuhul, or the Spume of Luna, or incombustible sulphur, or roasted calx, or alum which is out of apples, as ye know. But if there was any singular regimen for any of these, a Philosopher would not say so, as ye know. Understand, therefore, that sulphur, calx, and alum which is from apples, and Kuhul, are all nothing else but water of sulphur. Know ye also that Magnesia, being mixed with quicksilver and sulphur, they pursue one another. Hence you must not dismiss that Magnesia without the quicksilver, for when it is composed it is called an exceeding strong composition, which is one of the ten regimens established by the Philosophers. Know, also, that when Magnesia is whitened with quicksilver, you must congeal white water therein, but when it is reddened you must congeal red water, for, as the Philosophers have observed in their books, the regimen is not one. Accordingly, the first congelation is of tin, copper, and lead. But the second is composed with water of sulphur. Some, however, reading this book, think that the composition can be bought. It must be known for certain that nothing of the work can be bought, and that the science of this Art is nothing else than vapour and the sublimation of water, with the conjunction, also, of quicksilver in the body of Magnesia; but, heretofore, the Philosophers have demonstrated in their books that the impure water of sulphur is from sulphur only, and no sulphur is produced without the water of its calx, and of quicksilver, and of sulphur.

The Forty-Ninth Dictum.

Belus saith: O all ye Philosophers, ye have not dealt sparingly concerning composition and contact, but cornposition, contact, and congelation are one thing! Take, therefore, a part From the one composition and a part out of ferment of gold, and on these impose pure water of sulphur. This, then, is the potent (or revealed) arcanum which tinges every body.

Pythagoras answereth: O Belus, why hast thou called it a potent arcanum, yet hast not shown its work!

And he: In our books, O Master, we have found the same which thou hast received from the ancients!

And Pythagoras: Therefore have I assembled you together, that you might remove any obscurities which are in any books.

And he: Willingly, O Master! It is to be noted that pure water which is from sulphur is not composed of sulphur alone, but is composed of several things, for the one sulphur is made out of several sulphurs. How, therefore, O Master, shall I compose these things that they may become one!

And he: Mix, O Belus, that which strives with the fire with that which does not strive, for things which are conjoined in a fire suitable to the same contend, because the warm venoms of the physician are cooked in a gentle, incomburent fire! Surely ye perceive what the Philosophers have stated concerning decoction, that a little sulphur burns many strong things, and the humour which remains is called humid pitch, balsam of gum, and other like things. Therefore our Philosophers are made like to the physicians, notwithstanding that the tests of the physicians are more intense than those of the Philosophers.

The Turba answereth: I wish, O Belus, that you would also shew the disposition of this potent arcanum!

And he: I proclaim to future generations that this arcanum proceeds from two compositions, that is to say, sulphur and magnesia. But after it is reduced and conjoined into one, the Philosophers have called it water, spume of Boletus (i.e., a species of fungus), and the thickness of gold. When, however, it has been reduced into quicksilver, they call it sulphur of water; sulphur also, when it contains sulphur, they term a fiery venom, because it is a potent (or open) arcanum which ascends from those things ye know.

The Fiftieth Dictum.

Pandolphus saith: If, O Belus, thou dost describe the sublimation of sulphur for future generations, thou wilt accomplish an excellent thing!

And the Turba: Do thou show it forth, therefore, O Pandolphus!

And he: The philosophers have ordered that quicksilver should be taken out of Cambar, and albeit they spoke truly, yet in these words there is a little ambiguity, the obscurity of which I will remove. See then that the quicksilver is sublimed in tabernacles, and extract the same from Cambar, but there is another Cambar in sulphur which Belus hath demonstrated to you, for out of sulphur mixed with sulphur, many works proceed. When the same has been sublimed, there proceeds from the Cambar that quicksilver which is called Ethelia, Orpiment, Zendrio, or Sanderich, Ebsemich, Magnesia, Kuhul, or Chuhul, and many other names. Concerning this, philosophers have said that, being ruled by its regimen (for ten is the perfection of all things), its white nature appears, nor is there any shadow therein. Then the envious have called it lead from Ebmich, Magnesia, Marteck, White Copper. For, when truly whitened, it is devoid of shadow and

blackness, it has left its thickened ponderous
bodies, and therewith a clean humid spirit has
ascended, which spirit is tincture. Accordingly, the
wise have said that copper has a soul and a body.
Now, its soul is spirit, and its body is thick.
Therefore, it behoves you to destroy the thick body
until ye extract a tingeing spirit from the same.
Mix, also, the spirit extracted therefrom with light
sulphur until you, investigators, find your design
accomplished.

The Fifty-First Dictum.

Horfolcos saith: Thou hast narrated nothing, O Pandolphus, save the last regimen of this body! Thou hast, therefore, composed an ambiguous description for readers. But if its regimen were commenced from the beginning, you would destroy this obscurity.

Saith the Turba: Speak, therefore, concerning this to posterity, so far as it may please you.

And he: It behoves you, investigators of this Art, first to burn copper in a gentle fire, like that required in the hatching of eggs. For it behoves you to burn it with its humidity lest its spirit be burnt, and let the vessel be closed on all sides, so that its colour [heat] may be increased, the body of copper be destroyed, and its tingeing spirit be extracted, concerning which the envious have said: Take quicksilver out of the Flower of Copper, which also they have called the water of our copper, a fiery venom, and a substance extracted from all things, which further they have termed Ethelia, extracted out of many things. Again, some have said that when all things become one, bodies are made not-bodies, but not-bodies bodies. And know, all ye investigators of this Art, that every body is dissolved with the spirit with which it is mixed, with which without doubt it becomes a similar spiritual thing, and that every spirit which has a tingeing colour of spirits, and is constant against fire, is altered and coloured by bodies. Blessed then be the name of Him who hath inspired the Wise with the idea of turning a body into a spirit having strength and colour, unalterable and incorruptible, so that what formerly was volatile sulphur is now made sulphur not-volatile, and incombustible! Know, also, all ye sons of learning, that he who is able to make your fugitive spirit red by the body mixed with it, and then from that body and that spirit can extract the tenuous nature hidden in the belly thereof, by a most subtle regimen, tinges every body, if only he is patient in spite of the tedium of extracting. Wherefore the envious have said: Know that out of copper, after it is humectated by the moisture thereof, is pounded in its water, and is cooked in sulphur, if ye extract a body having Ethelia, ye will find that which is suitable as a tincture for anything. Therefore the envious have said: Things that are diligently pounded in the fire, with sublimation of the Ethelia, become fixed tinctures. For whatsoever words ye find in any man's book signify quicksilver, which we call water of sulphur, which also we sometimes say is lead and copper and copulated coin.

The Fifty-Second Dictum.

Ixumdrus saith: You will have treated most excellently, O Horfolcus, concerning the regimen of copper and the humid spirit, provided you proceed therewith.

And he: Perfect, therefore, what I have omitted, O Ixumdrus!

Ixumdrus saith: You must know that this Ethelia which you have previously mentioned and notified, which also the envious have called by many names, doth whiten, and tinge when it is whitened; then truly the Philosophers have called it the Flower of Gold, because it is a certain natural thing. Do you not remember what the Philosophers have said, that before it arrives at this terminus, copper does not tinge? But when it is tinged it tinges, because quicksilver tinges when it is combined with its tincture. But when it is mixed with those ten things which the Philosophers have denominated fermented urines, then have they called all these things Multiplication. But some have termed their mixed bodies Corsufle and Gum of Gold. Therefore, those names which are found in the books of the Philosophers, and are thought superfluous and vain, are true and yet are fictitious, because they are one thing, one opinion, and one way. This is the quicksilver which is indeed extracted from all things, out of which all things are produced, which also is pure water that destroys the shade of copper. And know ye that this quicksilver, when it is whitened, becomes a sulphur which contains sulphur, and is a venom that has a brilliance like marble; this the envious call Ethelia, orpiment and sandarac, out of which a tincture and pure spirit ascends with a mild fire, and the whole pure flower is sublimated, which flower becomes wholly quicksilver. It is, therefore, a most great arcanum which the Philosophers have thus described, because sulphur alone whitens copper. Ye, O investigators of this Art, must know that the said sulphur cannot whiten copper until it is whitened in the work! And know ye also that it is the habit of this sulphur to escape. When, therefore, it flees from its own thick bodies, and is sublimated as a vapour, then it behoves you to retain it otherwise with quicksilver of its own kind, lest it vanish altogether. Wherefore the Philosophers have said, that sulphurs are contained by sulphurs. Know, further, that sulphurs tinge, and then are they certain to escape unless they are united to quicksilver of its own kind. Do not, therefore, think that because it tinges and afterwards escapes, it is the coin of the Vulgar, for what the Philosophers are seeking is the

coin of the Philosophers, which, unless it be mixed
with white or red, which is quicksilver of its own
kind, would doubtless escape. I direct you,
therefore, to mix quicksilver with quicksilver (of
its kind) until together they become one clean water
composed out of two. This is, therefore, the great
arcanum, the confection of which is with its own
gum; it is cooked with flowers in a gentle fire and
with earth; it is made red with mucra and with
vinegar, salt, and nitre, and with mutal is turned
into rubigo, or by any of the select tingeing agents
existing in our coin.

The Fifty-Third Dictum.

Exumenus saith: The envious have laid waste the whole Art with the multiplicity of names, but the entire work must be the Art of the Coin. For the Philosophers have ordered the doctors of this art to make coin-like gold, which also the same Philosophers have called by all manner of names.

The Turba answereth: Inform, therefore, posterity, O Exumenus, concerning a few of these names, that they may take warning!

And he: They have named it salting, sublimating, washing, and pounding Ethelias, whitening in the fire, frequently cooking vapour and coagulating, turning into rubigo, the confection of Ethel, the art of the water of sulphur and coagula. By all these names is that operation called which has pounded and whitened copper. And know ye, that quicksilver is white to the sight, but when it is possessed by the smoke of sulphur, it reddens and becomes Cambar. Therefore, when quicksilver is cooked with its confections it is turned into red, and hence the Philosopher saith that the nature of lead is swiftly converted. Do you not see that the Philosophers have spoken without envy! Hence we deal in many ways with pounding and reiteration, that ye may extract the spirits existing in the vessel, which the fire did not cease to burn continuously. But the water placed with those things prevents the fire from burning, and it befalls those things that the more they are possessed by the flame of fire, the more they are hidden in the depths of the water, lest they should be injured by the heat of the fire; but the water receives them in its belly and repels the flame of fire from them.

The Turba answereth: Unless ye make bodies not-bodies ye achieve nothing. But concerning the sublimation of water the Philosophers have treated not a little. And know that unless ye diligently pound the thing in the fire, the Ethelia does not ascend, but when that does not ascend ye achieve nothing. When, however, it ascends it is an instrument for the intended tincture with which ye tinge, and concerning this Ethelia Hermes saith: Sift the things which ye know; but another: Liquefy the things. Therefore, Arras saith: Unless ye pound the thing diligently in the fire, Ethelia does not ascend. The Master hath put forth a view which I shall now explain to the reasoners. Know ye that a very great wind of the south, when it is stirred up, sublimates clouds and elevates the vapours of the sea.

The Turba answereth: Thou hast dealt obscurely.

And he: I will explain the testa, and the vessel wherein is incombustible sulphur. But I order you to congeal fluxible quicksilver out of many things, that two may be made three, and four one, and two one.

The Fifty-Fourth Dictum.

Anaxagoras saith: Take the volatile burnt thing which lacks a body, and incorporate it. Then take the ponderous thing, having smoke, and thirsting to imbibe.

The Turba answereth: Explain, O Anaxagoras, what is this obscurity which you expound, and beware of being envious!

And he: I testify to you that this volatile burnt thing, and this other which thirsts, are Ethelia, which has been conjoined with sulphur. Therefore, place these in a glass vessel over the fire, and cook until the whole becomes Cambar. Then God will accomplish the arcanum ye seek. But I direct you to cook continuously, and not to grow tired of repeating the process. And know ye that the perfection of this work is the confection of water of sulphur with tabula; finally, it is cooked until it becomes Rubigo, for all the Philosophers have said: He who is able to turn Rubigo into golden venom has already achieved the desired work, but otherwise his labour is vain.

The Fifty-Fifth Dictum.

Zenon saith: Pythagoras hath treated concerning the water, which the envious have called by all names. Finally, at the end of his book he has treated of the ferment of gold, ordaining that thereon should be imposed clean water of sulphur, and a small quantity of its gum. I am astonished, O all ye Turba, how the envious have in this work discoursed of the perfection rather than the commencement of the same!

The Turba answereth: Why, therefore, have you left it to putrefy?

And he: Thou hast spoken truly; putrefaction does not take place without the dry and the humid. But the vulgar putrefy with the humid. Thus the humid is merely coagulated with the dry. But out of both is the beginning of the work. Notwithstanding, the envious have divided this work into three parts, asserting that one quickly flees, but the other is fixed and immovable.

The Fifty-Sixth Dictum.

Constans saith: What have you to do with the treatises of the envious, for it is necessary that this work should deal with four things?

They answer: Demonstrate, therefore, what are those four?

And he: Earth, water, air, and fire. Ye have then those four elements without which nothing is ever generated, nor is anything absolved in the Art. Mix, therefore, the dry with the humid, which are earth and water, and cook in the fire and in the air, whence the spirit and the soul are dessicated. And know ye that the tenuous tingeing agent takes its power out of the tenuous part of the earth, out of the tenuous part of the fire and of the air, while out of the tenuous part of the water, a tenuous spirit has been dessicated. This, therefore, is the process of our work, namely, that everything may be turned into earth when the tenuous parts of these things are extracted, because a body is then composed which is a kind of atmospheric thing, and thereafter tinges the imposed body of coins. Beware, however, O all ye investigators of this art, lest ye multiply things, for the envious have multiplied and destroyed for you! They have also described various

regimens that they might deceive; they have further
called it (or have likened it to) the humid with all
the humid, and the dry with all the dry, by the name
of every stone and metal, gall of animals of the
sea, the winged things of heaven and reptiles of the
earth. But do ye who would tinge observe that bodies
are tinged with bodies. For I say to you what the
Philosopher said briefly and truly at the beginning
of his book. In the art of gold is the quicksilver
from Cambar, and in coins is the quicksilver from
the Male. In nothing, however, look beyond this,
since the two quicksilvers are also one.

The Fifty-Seventh Dictum.

Acratus saith: I signify to posterity that I make philosophy near to the Sun and Moon. He, therefore, that will attain to the truth let him take the moisture of the Sun and the Spume of the Moon.

The Turba answereth: Why are you made an adversary to your brethren?

And he: I have spoken nothing but the truth.

But they: Take what the Turba hath taken.

And he: I was so intending, yet, if you are willing, I direct posterity to take a part of the coins which the Philosophers have ordered, which also Hermes has adapted to the true tingeing, and a part of the copper of the Philosophers, to mix the same with the coins, and place all the four bodies in the vessel, the mouth of which must be carefully closed, lest the water escape. Cooking must proceed for seven days, when the copper, already pounded with the coins, is found turned into water. Let both be again slowly cooked, and fear nothing. Then let the vessel be opened, and a blackness will appear above. Repeat the process, cook continually until the blackness of Kuhul, which is from the blackness of coins, be consumed. For when that is consumed a precious whiteness will appear on them; finally, being returned to their place, they are cooked until the whole is dried and is turned into stone. Also repeatedly and continuously cook that stone born of copper and coins with a fire sharper than the former, until the stone is destroyed, broken up, and turned into cinder, which is a precious cinder. Alas, O ye sons of the Doctrine, how precious is that which is produced from it! Mixing, therefore, the cinder with water, cook again, until that cinder liquefy therewith, and then cook and imbue with permanent water, until the composition becomes sweet and mild and red. Imbue until it becomes humid. Cook in a still hotter fire, and carefully close the mouth of the vessel, for by this regimen fugitive bodies become not-fugitive, spirits are turned into bodies, bodies into spirits, and both are connected together. Then are spirits made bodies having a tingeing and germinating soul.

The Turba answereth: Now hast thou notified to posterity that Rubigo attaches itself to copper after the blackness is washed off with permanent water. Then it is congealed and becomes a body of Magnesia. Finally, it is cooked until the whole body is broken up. Afterwards the volatile is turned into a cinder and becomes copper without its shadow. Attrition also truly takes place. Concerning, therefore, the work of the Philosophers, what hast thou delivered to posterity, seeing that thou hast by no means called things by their proper names!

And he: Following your own footsteps, I have discoursed even as have you.

Bonellus answereth: You speak truly, for if you did otherwise we should not order your sayings to be written in our books.

The Fifty-Eighth Dictum.

Balgus saith: The whole Turba, O Acratus, has already spoken, as you have seen, but a benefactor sometimes deceives, though his intention is to do good.

And they: Thou speakest truly. Proceed, therefore, according to thy opinion, and beware of envy!

Then he: You must know that the envious have described this arcanum in the shade; in physical reasoning and astronomy, and the art of images; they have also likened it to trees; they have ambiguously concealed it by the names of metals, vapours, and reptiles; as is generally perceived in all their work. I, nevertheless, direct you, investigators of this science, to take iron and draw it into plates; finally, mix (or sprinkle) it with venom, and place it in its vessel, the mouth of which must be closed most carefully, and beware lest ye too much increase the humour, or, on the other hand, lest it be too dry, but stir it vigorously as a mass, because, if the water be in excess, it will not be contained in the chimney, while, if it be too dry, it will neither be conjoined nor cooked in the chimney; hence I direct you to confect it diligently; finally, place it in its vessel, the mouth of which must be closed internally and externally with clay, and, having kindled coals above it, after some days ye shall open it, and there shall ye find the iron plates already liquefied; while on the lid of the vessel ye shall find globules. For when the fire is kindled the vinegar ascends, because its spiritual nature passes into the air, wherefore, I direct you to keep that part separately. Ye must also know that by multiplied decoctions and attritions it is congealed and coloured by the fire, and its nature is changed. By a similar decoction and liquefaction Cambar is not disjoined. I notify to you that by the said frequent decoction the weight of a third part of the water is consumed, but the residue becomes a wind in the Cambar of the second spirit. And know ye that nothing is more precious or more excellent than the red sand of the sea, for the Sputum of Luna is united with the light of the Sun's rays. Luna is perfected by the coming on of night, and by the heat of the Sun the dew is congealed. Then, that being wounded, the dew of the death dealer is joined, and the more the days pass on the more intensely is it congealed, and is not burned. For he who cooks with the Sun is himself congealed, and that signal whiteness causes it to overcome the terrene fire.

Then saith Bonites: Do you not know, O Balgus, that the Spume of Luna tinges nothing except our copper?

And Balgus: Thou speakest truly.

And he: Why, therefore, hast thou omitted to describe that tree, of the fruit whereof whosoever eateth shall hunger nevermore?

And Balgus: A certain person, who has followed science, has notified to me after what manner he discovered this same tree, and appropriately operating, did extract the fruit and eat of it. But when I inquired of him concerning the growth and the increment, he described that pure whiteness, thinking that the same is found without any laborious disposition. Then its Perfection is the fruit thereof. But when I further asked how it is nourished with food until it fructifies, he said: Take that tree, and build a house about it, which shall wholly surround the same, which shall also be circular, dark, encircled by dew, and shall have placed on it a man of a hundred years; shut and secure the door lest dust or wind should reach them. Then in the time of 180 days send them away to their homes. I say that man shall not cease to eat of the fruit of that tree to the perfection of the number [of the days] until the old man shall become young. O what marvellous natures, which have transformed the soul of that old man into a juvenile body, and the father is made into the son! Blessed be thou, O most excellent God!

The Fifty-Ninth Dictum.

Theophilus saith: I propose to speak further concerning those things which Bonites hath narrated.

And the Turba: Speak, Brother, for thy brother hath discoursed elegantly.

And he: Following in the steps of Bonites I will make perfect his sayings. It should be known that all the Philosophers, while they have concealed this disposition, yet spoke the truth in their treatises when they named water of life, for this reason, that whatsoever is mixed with the said water first dies, then lives and becomes young. And know, all ye disciples, that iron does not become rusty except by reason of this water, because it tinges the plates; it is then placed in the sun till it liquefies and is imbued, after which it is congealed. In these days it becomes rusty, but silence is better than this illumination.

The Turba answereth: O Theophilus, beware of becoming envious, and complete thy speech!

And he: Would that I might repeat the like thing!

And they: What is thy will?

Then he: Certain fruits, which proceed first from that perfect tree, do flourish in the beginning of the summer, and the more they are multiplied the more are they adorned, until they are perfected, and being mature become sweet. In the same way that woman, fleeing from her own children, with whom she lives, although partly angry, yet does not brook being overcome, nor that her husband should possess her beauty, who furiously loves her, and keeps awake contending with her, till he shall have carnal intercourse with her, and God make perfect the foetus, when he multiplies children to himself according to his pleasure. His beauty, therefore, is consumed by fire who does not approach his wife except by reason of lust. For when the term is finished he turns to her. I also make known to you that the dragon never dies, but the Philosophers have put to death the woman who slays her spouses. For the belly of that woman is full of weapons and venom. Let, therefore, a sepulchre be dug for the dragon, and let that woman be buried with him, who being strongly joined with that woman, the more he clasps her and is entwined with her, the more his body, by the creation of female weapons in the body of the woman, is cut up into parts. For perceiving him mixed with the limbs of a woman he becomes secure from death, and the whole is turned into blood. But the Philosophers, beholding him turned

into blood, leave him in the sun for certain days,
until the lenitude is consumed, the blood dries up,
and they find that venom which now is manifest. Then
the wind is hidden.

The Sixtieth Dictum.

Bonellus saith: Know, all ye disciples, that out of the elect things nothing becomes useful without conjunction and regimen, because sperma is generated out of blood and desire. For the man mingling with the woman, the sperm is nourished by the humour of the womb, and by the moistening blood, and by heat, and when forty nights have elapsed the sperm is formed. But if the humidity of the blood and of the womb were not heat, the sperm would not be dissolved, nor the foetus be procreated. But God has constituted that heat and blood for the nourishment of the sperm until the foetus is brought forth, after which it is not nourished, save by milk and fire, sparingly and gradually, while it is dust, and the more it burns the more, the bones being strengthened, it is led towards youth, arriving at which it is independent. Thus it behoves you also to act in this Art. Know ye that without heat nothing is ever generated, and that the bath causes the matter to perish by means of intense heat. If, indeed, it be frigid, it puts to flight and disperses, but if it have been tempered, it is convenient and sweet to the body, wherefore the veins become smooth and the flesh is augmented. Behold it has been demonstrated to you, all ye disciples! Understand, therefore, and in all things which ye attempt to rule, fear God.

The Sixty-First Dictum.

Moses saith: It is to be observed that the envious have named lead of copper instruments of formation, simulating, deceiving posterity, to whom I give notice that there are no instruments except from our own white, strong, and splendid powder, and from our concave stone and marble, to the whole work whereof there is no more suitable powder, nor one more conjoined to our composition, than the powder of Alociae, out of which are produced instruments of formation. Further, the Philosophers have already said: Take instruments out of the egg. Yet they have not said what the egg is, nor of what bird. And know ye that the regimen of these things is more difficult than the entire work, because, if the composition be ruled more than it should be, its light is taken and extinguished by the sea. Wherefore the Philosophers have ordered that it should be ruled with profound judgment. The moon, therefore, being at the full, take this and place in sand till it be dissolved. And know ye that while ye are placing the same in sand and repeating the process, unless ye have patience, ye err in ruling, and corrupt the work. Cook, therefore, the same in a gentle fire until ye see that it is dissolved. Then extinguish with vinegar, and ye shall find one thing separated from three companions. And know ye that the first, Ixir, commingles, the second burns, while the third liquefies. In the first place, therefore, impose nine ounces of vinegar twice - first while the vessel is being made hot, and second when it is heated.

The Sixty-Second Dictum.

Mundus saith: It behoves you, O all ye seekers after this Art, to know that whatsoever the Philosophers have narrated or ordained, Kenckel, herbs, geldum, and carmen, are one thing! Do not, therefore, trouble about a plurality of things, for there is one Tyrian tincture of the Philosophers to which they have given names at will, and having abolished the proper name, they have called it black, because it has been extracted from our sea. And know that the ancient priests did not condescend to wear artificial garments, whence, for purifying altars, and lest they should introduce into them anything sordid or impure, they tinged Kenckel with a Tyrian colour; but our Tyrian colour, which they placed in their altars and treasuries, was more clean and fragrant than can be described by me, which also has been extracted from our red and most pure sea, which is sweet and of a pleasant odour, and is neither sordid nor impure in putrefaction. And know ye that we have given many names to it. which are all true – an example of which, for those that possess understanding, is to be traced in corn that is being ground. For after grinding it is called by another name, and after it has been passed through the sieve, and the various substances have been separated one from another, each of these has its own name, and yet fundamentally there is but one name, to wit, corn, from which many names are distinguished. Thus we call the purple in each grade of its regimen by the name of its own colour.

The Sixty-Third Dictum.

Philosophus saith: I notify to posterity that the nature is male and female, wherefore the envious have called it the body of Magnesia, because therein is the most great arcanum! Accordingly, O all ye seekers after this Art, place Magnesia in its vessel, and cook diligently! Then, opening it after some days, ye shall find the whole changed into water. Cook further until it be coagulated, and contain itself. But, when ye hear of the sea in the books of the envious, know that they signify humour, while by the basket they signify the vessel, and by the medicines they mean Nature, because it germinates and flowers. But when the envious say: Wash until the blackness of the copper passes away, certain people name this blackness coins. But Agadimon has clearly demonstrated when he boldly put forth these words: It is to be noted, O all ye demonstrators of this art, that the things [or the copper] being first mixed and cooked once, ye shall find the prescribed blackness! That is to say, they all become black. This, therefore, is the lead of the Wise, concerning which they have treated very frequently in their books. Some also call it [the lead] of our black coins.

The Sixty-Fourth Dictum.

Pythagoras saith: How marvellous is the diversity of the Philosophers in those things which they formerly asserted, and in their coming together [or agreement], in respect of this small and most common thing, wherein the precious thing is concealed! And if the vulgar knew, O all ye investigators of this art, the same small and vile thing, they would deem it a lie! Yet, if they knew its efficacy, they would not vilify it, but God hath concealed this from the crowd lest the world should be devastated.

The Sixty-Fifth Dictum.

Horfolcus saith: You must know, O all ye who love wisdom, that whereas Mundus hath been teaching this Art, and placing before you most lucid syllogisms, he that does not understand what he has said is a brute animal! But I will explain the regimen of this small thing, in order that any one, being introduced into this Art, may become bolder, may, more assuredly consider it, and although it be small, may compose the common with that which is dear, and the dear with that which is common. Know ye that in the beginning of the mixing, it behoves you to commingle elements which are crude, gentle, sincere, and not cooked or governed, over a gentle fire. Beware of intensifying the fire until the elements are conjoined, for these should follow one another, and be embraced in a complexion, whereby they are gradually burnt, until they be dessicated in the said gentle fire. And know that one spirit burns one thing and destroys one thing, and one body strengthens one spirit, and teaches the same to contend with the fire. But, after the first combustion, it is necessary that it should be washed, cleansed, and dealbated on the fire until all things become one colour; with which, afterwards, it behoves you to mix the residuum of the whole humour, and then its colour will be exalted. For the elements, being diligently cooked in the fire, rejoice, and are changed into different natures, because the liquefied, which is the lead, becomes not-liquefied, the humid becomes dry, the thick body becomes a spirit, and the fleeing spirit becomes strong and fit to do battle against the fire. Whence the Philosopher saith: Convert the elements and thou shalt find what thou seekest. But to convert the elements is to make the moist dry and the fugitive fixed. These things being accomplished by the disposition, let the operator leave it in the fire until the gross be made subtle, and the subtle remain as a tingeing spirit. Know ye, also, that the death and life of the elements proceed from fire, and that the composite germinates itself, and produces that which ye desire, God favouring. But when the colours begin ye shall behold the miracles of the wisdom of God, until the Tyrian colour be accomplished. O wonder-working Nature, tingeing other natures! O heavenly Nature, separating and converting the elements by regimen! Nothing, therefore, is more precious than these Natures in that Nature which multiplies the composite, and makes fixed and scarlet.

The Sixty-Sixth Dictum.

Exemiganus saith: Thou hast already treated, O Lucas, concerning living and concealed silver, which is Magnesia, as it behoves thee, and thou hast commanded posterity to prove [or to experiment] and to read the books, knowing what the Philosophers have said: Search the latent spirit and disesteem it not, seeing that when it remains it is a great arcanum and effects many good things.

The Sixty-Seventh Dictum.

Lucas saith: I testify to posterity, and what I set forth is more lucid than are your words, that the Philosopher saith: Burn the copper, burn the silver, burn the gold.

Hermiganus replies: Behold something more dark than ever!

The Turba answereth: Illumine, therefore, that which is dark.

And he: As to that which he said - Burn, burn, burn, the diversity is only in the names, for they are one and the same thing.

And they: Woe unto you! how shortly hast thou dealt with it! why art thou Poisoned with jealousy!

And he: Is it desirable that I should speak more clearly?

And they: Do so.

And he: I signify that to whiten is to burn, but to make red is life. For the envious have multiplied many names that they might lead posterity astray, to whom I testify that the definition of this Art is the liquefaction of the body and the separation of the soul from the body, seeing that copper, like a man, has a soul and a body. Therefore, it behoves you, O all ye Sons of the Doctrine, to destroy the body and extract the soul therefrom! Wherefore the Philosophers said that the body does not penetrate the body, but that there is a subtle nature, which is the soul, and it is this which tinges and penetrates the body. In nature, therefore, there is a body and there is a soul.

The Turba answereth: Despite your desire to explain, you have put forth dark words.

And he: I signify that the envious have narrated and said that the splendour of Saturn does not appear unless it perchance be dark when it ascends in the air, that Mercury is hidden by the rays of the Sun, that quicksilver vivifies the body by its fiery strength, and thus the work is accomplished. But Venus, when she becomes oriental, precedes the Sun.

The Sixty-Eighth Dictum.

Attamus saith: Know, O all ye investigators of this Art, that our work, of which ye have been inquiring, is produced by the generation of the sea, by which and with which, after God, the work is completed! Take, therefore, Halsut and old sea stones, and boil with coals until they become white. Then extinguish in white vinegar. If 24 ounces thereof have been boiled, let the heat be extinguished with a third part of the vinegar, that is, 8 ounces; pound with white vinegar, and cook in the sun and black earth for 42 days. But the second work is performed from the tenth day of the month of September to the tenth day [or grade] of Libra. Do not impose the vinegar a second time in this work, but leave the same to be cooked until all its vinegar be dried up and it becomes a fixed earth, like Egyptian earth. And the fact that one work is congealed more quickly and another more slowly, arises from the diversity of cooking. But if the place where it is cooked be humid and dewy it is congealed more quickly, while if it be dry it is congealed more slowly.

The Sixty-Ninth Dictum.

Florus saith: I am thinking of perfecting thy treatise, O Mundus, for thou has not accomplished the disposition of the cooking!

And he: Proceed, O Philosopher!

And Florus: I teach you, O Sons of the Doctrine, that the sign of the goodness of the first decoction is the extraction of its redness!

And he: Describe what is redness.

And Florus: When ye see that the matter is entirely black, know that whiteness has been hidden in the belly of that blackness. Then it behoves you to extract that whiteness most subtly from that blackness, for ye know how to discern between them. But in the second decoction let that whiteness be placed in a vessel with its instruments, and let it be cooked gently until it become completely white. But when, O all ye seekers after this Art, ye shall perceive that whiteness appear and flowing over all, be certain that redness is hid in that whiteness! However, it does not behove you to extract it, but rather to cook it until the whole become a most deep red, with which nothing can compare. Know also that the first blackness is produced out of the nature of

Marteck, and that redness is extracted from that blackness, which red has improved the black, and has made peace between the fugitive and the non-fugitive, reducing the two into one.

The Turba answereth: And why was this?

And he: Because the cruciated matter when it is submerged in the body, changes it into an unalterable and indelible nature. It behoves you, therefore, to know this sulphur which blackens the body. And know ye that the same sulphur cannot be handled, but it cruciates and tinges. And the sulphur which blackens is that which does not open the door to the fugitive and turns into the fugitive with the fugitive. Do you not see that the cruciating does not cruciate with harm or corruption, but by co-adunation and utility of things? For if its victim were noxious and inconvenient, it would not be embraced thereby until its colours were extracted from it unalterable and indelible. This we have called water of sulphur, which water we have prepared for the red tinctures; for the rest it does not blacken; but that which does blacken, and this does not come to pass without blackness, I have testified to be the key of the work.

The Seventieth Dictum.

Mundus saith: Know, all ye investigators of this Art, that the head is all things, which if it hath not, all that it imposes profits nothing. Accordingly, the Masters have said that what is perfected is one, and a diversity of natures does not improve that thing, but one and a suitable nature, which it behoves you to rule carefully, for by ignorance of ruling some have erred. Do not heed, therefore, the plurality of these compositions, nor those things which the philosophers have enumerated in their books. For the nature of truth is one, and the followers of Nature have termed it that one thing in the belly whereof is concealed the natural arcanum. This arcanum is neither seen nor known except by the Wise. He, therefore, who knows how to extract its complexion and rules equably, for him shall a nature rise forth therefrom which shall conquer all natures, and then shall that word be fulfilled which was written by the Masters, namely, that Nature rejoices in Nature, Nature overcomes Nature, and Nature contains Nature; at the same time there are not many or diverse Natures, but one having in itself its own natures and properties, by which it prevails over other things. Do you not see that the Master has begun with one and finished one?

Hence has he called those unities Sulphureous Water,
conquering all Nature.

The Seventy-First Dictum.

Bracus saith: How elegantly Mundus hath described this sulphureous water! For unless solid bodies are destroyed by a nature wanting a body, until the bodies become not-bodies, and even as a most tenuous spirit, ye cannot [attain] that most tenuous and tingeing soul, which is hidden in the natural belly. And know that unless the body be withered up and so destroyed that it dies, and unless ye extract from it its soul, which is a tingeing spirit, ye are unable to tinge a body therewith.

The Seventy-Second Dictum.

Philosophus saith: The first composition, that is, the body of Magnesia, is made out of several things, although they become one, and are called by one name, which the ancients have termed Albar of copper. But when it is ruled it is called by ten names, taken from the colours which appear in the regimen of the body of this Magnesia. It is necessary, therefore, that the lead be turned into blackness; then the ten aforesaid shall appear in the ferment of gold, with sericon, which is a composition called by ten names. When all these things have been said, we mean nothing more by these names than Albar of copper, because it tinges every body which has entered into the composition. But composition is twofold - one is humid, the other is dry. When they are cooked prudently they become one, and are called the good thing of several names. But when it becomes red it is called Flower of Gold, Ferment of Gold, Gold of Coral, Gold of the Beak. It is also called redundant red sulphur and red orpiment. But while it remains crude lead of copper, it is called bars and plates of metal. Behold I have revealed its names when it is raw, which also we should distinguish from the names when it has been cooked. Let it therefore be pondered over. It

behoves me now to exhibit to you the quantity of the
fire, and the numbers of its days, and the diversity
of intensity thereof in every grade, so that he who
shall possess this book may belong unto himself, and
be freed from poverty, so that he shall remain
secure in that middle way which is closed to those
who are deficient in this most precious art. I have
seen, therefore, many kinds of fire. One is made out
of straw and cinder, coals and flame, but one
without flame. Experiment shows that there are
intermediate grades between these kinds. But lead is
lead of copper, in which is the whole arcanum. Now,
concerning the days of the night in which will be
the perfection of the most great arcanum, I will
treat in its Proper place in what follows. And know
most assuredly that if a little gold be placed in
the composition, there will result a patent and
white tincture. Wherefore also a sublime gold and a
patent gold is found in the treasuries of the former
philosophers. Wherefore those things are unequal
which they introduce into their composition.
Inasmuch as the elements are commingled and are
turned into lead of copper, coming out of their own
former natures, they are turned into a new nature.
Then they are called one nature and one genus. These
things being accomplished, it is placed in a glass
vessel, unless in a certain way the composition
drinks the water and is altered in its colours. In

every grade it is beheld, when it is coloured by a venerable redness. Although concerning this elixir we read in the sayings of the philosophers: Take gold, occurring frequently, it is only needful to do so once. Wishing, therefore, to know the certitude of the adversary, consider what Democritus saith, how he begins speaking from bottom to top, then reversing matters he proceeds from top to bottom. For, he said: Take iron, lead, and albar for copper, which reversing, he again says: And our copper for coins, lead for gold, gold for gold of coral, and gold of coral for gold of crocus. Again, in the second place, when he begins from the top to the bottom, he saith: Take gold, coin, copper, lead, and iron; he shews, therefore, by his sayings that only semi-gold is taken. And without doubt gold is not changed into rust without lead and copper, and unless it be imbued with vinegar known by the wise, until, being cooked, it is turned into redness. This, therefore, is the redness which all the Philosophers signified, because, however they said: Take gold and it becomes gold of coral; Take gold of coral and it becomes purple gold - all these things are only names of those colours, for it behoves them that vinegar be placed in it, because these colours come from it. But by these things which the Philosophers have mentioned under various names, they have signified stronger bodies and forces. It is taken, therefore, once, that it may become rubigo and then vinegar is imposed on it. For when the said colours appear, it is necessary that each be decocted in forty days, so that it may be desiccated, the water being consumed; finally being imbued and placed in the vessel, it is cooked until its utility appear. Its first grade becomes as a citrine mucra, the second as red, the third as the dry pounded crocus of the vulgar. So is it imposed upon coin.

Conclusion.

Agmon saith: I will add the following by way of a corollary. Whosoever does not liquefy and coagulate errs greatly. Therefore, make the earth black; separate the soul and the water thereof, afterwards whiten; so shall ye find what ye seek. I say unto you that whoso makes earth black and then dissolves with fire, till it becomes even like unto a naked sword, who also fixes the whole with consuming fire, deserves to be called happy, and shall be exalted above the circle of the world. This much concerning the revelation of our stone, is, we doubt not, enough for the Sons of the Doctrine. The strength thereof, shall never become corrupted, but the same, when it is placed in the fire, shall be increased. If you seek to dissolve, it shall be dissolved; but if you would coagulate, it shall be coagulated. Behold, no one is without it, and yet all do need it! There are many names given to it, and yet it is called by one only, while, if need be, it is concealed. It is also a stone and not a stone, spirit, soul, and body; it is white, volatile, concave, hairless, cold, and yet no one can apply the tongue with impunity to its surface. If you wish that it should fly, it flies; if you say that it is water, you Speak the truth; if you say that it is not water, you speak falsely. Do not then be deceived by the multiplicity of names, but rest assured that it is one thing, unto which nothing alien is added. Investigate the place thereof, and add nothing that is foreign. Unless the names were multiplied, so that the vulgar might be deceived, many would deride our wisdom.

-Finis-

middle age, but indicated he was much older and the portrait clearly represented him as he looked more than a century earlier.

Contained therein, are letters between Gualdus and Baron von Reusenstein. The latter refers to Gualdus as an Adept and Dr Sigismund Bacstrom held von Reusenstein in very high regard. We have here, also, Gualdus' recipe for longevity. A subject, it seems, he knew quite a lot about!

Introduction

When the Almighty Creator created man immortal, he planted at the same time the Tree of Life in the center of the earthly paradise, as we read in the First Book of Moses, so that its fruit should be an antidote and Universal Medicine for man and serve in all cases of adverse happenings. But as Adam has deprived us of this benefit by his sin, the only thing left to man is the desire to live long, which is also considered the noblest among all transitory goods. Just as the Lord God Himself wanting to prompt the children all the more to obedience to their parents, has attached the promise of a long life to the Fourth Commandment (in the Fourth Book of Moses): So that you may live long and prosper on earth.

Even so, we must not think of achieving immortality through the Universal Medicine as we could have done through the fruits of the Tree of Life. Thus there are among all men only two, Enoch and Elias, who did not die but, as the Scripture relates, were transported into Paradise, although some believe, yes, the Apostles themselves, that the identical privilege of not tasting death had been promised to St John, Christ's favorite disciple, in the words

176

found in his Gospel, where the Saviour replies to Peter: If I wish that he (St John) should stay here till I return, of what concern is that to you? To say it briefly: There is no one weary of life, but everyone is eager to see how he can add to his years. This is a gift which we can hope to obtain through the Universal medicine, whose power extends over the three natural realms: animal, vegetable, and mineral. Galen has given us a fine example of maintaining one's life a long time, by doing it himself for 140 years without ever suffering any indisposition, as he reports in his book on the art of living a long life.

Because the desire to live long and healthy is naturally common to all, many a man will undoubtedly remember and ask me what I think of that man of whom the Dutch newspaper wrote and assured us that he Had passed through Venice and had there stated unquestionably the he had really reached 400 years of age. Now one could rightly ask me if this was due to the Universal Medicine which keeps the Humidum Radicale (radical moisture) and the natural warmth in perfect balance, counteracts the debilities of old age, and frequently rejuvenates man. To this I will give my answer in Part 3.

In Part I, I will show that individuals at all times existed who had been living for a few centuries. In the second I will speak about things that are both within and without us human beings and may contribute toward extending our lifespan. In the third part I will give some useful and curious observations concerning the Universal Medicine and its preparation.

Although we are born to die and Tertullian says that God did not make man mortal after the Fall through anger but through great compassion, the Holy Scripture nevertheless shows us that in general man reached 700 years before the Fall. Adam lived for 930 years, Seth for 912, Cain for 910, and thus the lifespan was decreased by and by, till God fixed it at 120 ordinary years after the Fall. Nevertheless, Arphaxad, who was born 200 years after the Flood, lived for 300 years and his son Salem for 430. Heber, a son of Salem's from whom the Hebrews have their name, lived for 465 years.

But here one might think that these years were not like our years and were not solar buyt lunar years of 29 to 30 days, or that every season constituted a year, as was the case with the Chaldeans and Arcadians, according to the report of Lactantius. Or at most they might not have been more than the time

when the sun passes from one tropic to another, so
that, calculated against our years, theirs were only
half as long. However, they could not have been
lunar years, as otherwise the consequence would be
that many persons of our present time were living
longer than our forefathers, when 100 of our years
would make 1200 lunar years. At least, those years
consisted of 12 months, because Moses says in Book
7:2, speaking of the Flood, that Noah was 600 years
old when the Flood occurred on the 17th day of the
second Moon; and in Chapter 8:4, he says that the
Ark rested on a mountain in Armenia on the 27th day
of the 7th month; and that the peaks of the high
mountains began to appear in the water on the first
day of the 6th month. In addition he says in the
13th verse that Noh opened the Ark on the first day
of the first month in his 600th year. From this it
is evident that Moses counted the years in 12
months, and even if there were a difference in the
calculation of the months according to our year and
months, it would not amount to more than 12 days.

Relating to secular history, Homer informs us that
Prince Nestor, son of Neli, was nearly 300 years old
when he came to the aid of the Greeks against the
Trojans. Anacreon assures us that Arganthemius, king
of the Tartessos was 150 years; Cinirus, king of
Cyprus, lived to 100 years; and Aeginius, to 200

years. Petrus Maffeus, in his Indian history, testified that in Bengal an ordinary an was found, who had neither science nor erudition, who was 355 years old, and as proof thereof he related everything from memory that had happened in his lifetime, and which was quite in accord with the chronicles.

The great Seneca, a Spaniard by birth, reached 44 years of age and would without doubt have become much older if his cruel and unjust disciple Nero had not cut the thread of his life. Under the Emperor Trajan, Simon Cleophe, Bishop of Jerusalem, was crucified at the age of 120 years old; St Anthony, Bishop of Egypt was 150; and Cronius, his collaborator, 155. Emperor Claudius, after sufficiently examining the age of the Bolognese Titus Fullonius, finally found that he had been 150 years old. The Hun King Attila died in the 124th year of his life. Petrus de Natalibus asserts that St Severinus, Bishop of Tongres, reached 375 years of age and was ordained Bishop in his 197th year. Nicolaus de Comitatibus certifies that among the Bracmanniers there had been one of 300 years.

It is also easy for Nature to give to one man alone so many years as she might otherwise gives to many together, just as she bestows on a giant so much strength and matter as would at other times be required for the formation of many human bodies. Such a one was a man of the lineage of Turgan, not far from Lake Constance, who fought against the Saxons under Charlemagne. He impaled eight of them on his pike, carried them over his shoulder and walked through the Rhine with them to his comrades, saying: Look here the German frogs that I have caught; I cannot understand their croaking.

Guido Bonatus assures us that in 1223 he knew someone by the name of Richard, who was already 400 years old and could undeniably prove that he had seen service under Charlemagne. In addition, there is also much talk about a certain Jean du Temps who served in the war under Charlemagne and died under Louis VII in the year 1146. From this we may conclude that he must have lived close to 360 years, as Charlemagne had already been crowned Emperor in the year 800.

A certain Englishman was 152 years old, And just as I do not report anything without good reason, I also say here that the learned Monsieur Hubin, Royal French Treasurer, had given me that man's portrait, which he had received from M. Jacques du Perron, nephew of the Cardinal of the same name, Bishop of Angouleme and subsequently Euvreux, where he died as

Chief Almoner of the daughter of the King of England, Henry IV.

That Englishman was rather tall and heavy, and his name was Thomas Parck, of Winningthon, parish of Alverburg, County of Shropshire. He was born in 1433 and had already 152 years when he was put before King Charles I in England on 9 October 1635. He had seen nine Kings on the English throne: Edward IV, Richard III, Henry VII, Henry VIII, Elizabeth, James VI, and Charles I, father of the present ruling King. The good man praised God among other things also for the fact that he had always steadfastly adhered to the Roman Catholic religion, although he had seen three changes in religion during his lifetime, under Henry VIII, Queen Mary Tudor, and then again under Elizabeth. He readily confessed that he had been put on trial in his 100th year because he had got a young girl with child. This was also the reason why, after the customs of the country, he had to stand in front of the church door, covered with a white cloth and holding a wax candle in his hand, thus obviously doing penance. Sixteen years before his death he lost his sight. He died in London on 24 November 1635, within half an hour, without having been sich before or noticing the approaching death. After his death, his body was opened and all his inner parts were found to be healthy, except his lung which was full of hardened blood. The physicians attributed this to a change of air, as he had been brought to his place of death from a locality where the air had been much purer and milder than in London.

At the same time, the Countess of Arundel presented to the Queen of England a midwife who was 123 years old and still performed her duties in her home town two years before.

Olaus Magnus relates in his historical accounts that an English Bishop by name of David reached 170 years. Buchanan asserts that Laurence Hutland was still preaching in his 140th year, even in the severest of winters.

It is therefore evident from the ecclesiastical and temporal history that at all times the life of some persons lasted for a hundred years and that it has not always been so precisely limited to 60 and 80 years as it is said in the Psalms of David. Solomon also says that God did not make death, which signified nothing but a mere name without essence and only the absence of life. The said wise King also says that old age is a crown of honor and that grey hair is worth esteeming, because those who have it are generally useful and necessary to the community on account of their long experience.

Now we will also show which things that are outside of us and how many of them, such as the place of residence, purity of the air and water, etc., contribute to the preservation of health and the prolongation of life. It is rightly said that the dead are the best teachers because they teach us in their works without flattery and self-interest, as the following epitaph informs us in regard to longevity: Continence and Frugality Prolong Our Lifespan,

The great Pythagorean Appolonius of Tyana retained his youth for over 100 years by his continence and frugality, which also caused the ancient anchorites to extend their lifespan and health so far. Due precisely to continence and frugality, the great philosopher Democritus also enjoyed the best health for 109 years. And it is worth remembering what Diogenes Laertius thinks of his death: His sister having indicated that she could not attend the festivities of the goddess Ceres if he died before them, he complied and sustained himself during his last three days of life with the smell of warm bread. The true celibate and chaste state is indeed an excellent means for living long and healthy.

But Artaxerxes, King of Persia, who begot 115 children, ended his life only after 100 years, due solely to the conspiration of fifty of his children. The Roman Emperor Proclus once boasted that a hundred Polish girls had given birth to 100 of his children within two weeks.

In such a way, at the time of St Jerome, a Roman man could have had a legion of legitimate children with a Roman lady, according toe an account of this Church Father who lived at the time of Pope Damasu. In Rome there lived a man who had already become a widower twenty times. He had married a widow who, in turn, had already 20 husbands. When she died, he attended the corpse with a laurel wreath on his head and a palm branch in his hand, under the great cheers of other men, which they uttered because he had survived his otherwise incomparable wife.

Moderation and physical exercise also make us healthy and strong. This is why the Romans were extremely amazed at the strength and giant size of our Gauls, who did not drink wine and for a long time knew nothing about it, until the Swiss Helicon first brought the vine and its juice to Gaul. Likewise, the kind of food we eat contributes a great deal to a long life. People in the Province of

Limoges mostly eat nothing but chestnuts and live for a very long time, because they get from them nourishment that is little subject to corruption and thus does not easily dissipate.

A good constitution and the right temperature of the radical moisture and the natural warmth are necessary for a long life. The superfluous moisture tones the warmth down; and frequent warmth in turn tones down the moisture. That is why a long life can be expected of a sanguine constitution, as the blood is then warm and moist. The aggressive and ever active fire of the choleric constitution cannot last long in a dry person. The great moisture of the phlegmatic cannot sufficiently cook the natural warmth; and the melancholic is all too dry and cold. But it may well be that when the choleric and the phlegmatic constitutions are combined, and one tempers the other's faults, they make for a long life. This also happens with the sanguine and melancholic, when the warmth and moisture of the blood is tempered by the dryness of the melancholic and can likewise give a long life.

In addition, a healthy place of residence contributes not a little to a long life. When Emperor Vespasian and his son Titus had a census made of all of Italy, four men were found in a town of Vellajacum, region of Piacenza, who were each 210 years old, and still six other men each 110 years. At the same time, there lived in Arimini a woman by name of Tertulla who was 137 years old. Likewise another in Faventia in her 132nd year. Pliny relates from the Isigon that in India the Cirneses generally live to 140 years. Pomponius Mela reports that the residents of a town situated at the foot of Mount Atlas live twice as long as other inhabitants of the globe. And Orisicrates asserts that there are Indians in the tropics who are as tall as five elbows and live to 130 without debilitating old age. Ctesias likewise avers that those who lived in Pandoria in the valleys generally reached 200 years, and they had this odd characteristic that they had white and grey hair in their youth but black hair in old age.

Helanicus writes that in the region of Etolia the inhabitants usually became 200 years old. Among them, Pictoreus reached 300 years, according to a report by Damasus. If Ephorus is to be believed, the Arcadian King lived for 300 years. And Alexander Cornelius says that in Illyria a certain Dandon extended his life to 500 years. But in his Periplus Xenophon goes even further, saying that the Maritime King lived for 600 years and his son for 800.

Olaus Magnus writes in the 4th book of his Histories that the people of the coldest midnight countries usually live to 160 years of age; and in his 12th book he says that the inhabitants of Iceland reach more than 100 years.

In his Historia Naturali, Nierembergius asserts that the people living in the Jucantic mountains became quite old. And in the region of Versin in Brazil, as is testified to by Antonius Pipafelta, people very often reach 140 years. According to a report by Ludovic Bartama the age of 100 years is quite common in blessed Arabia.

In the Auvergne it often happens that fathers see the children of their grandchildren, and I have read somewhere that in our Alps a single man was the head and progenitor of a whole village which had more than a hundred households, all stemmed from him. Finally, in the year 1660 traveled from Ternant towards Orange, together with the Marquis of St Andre Montbrun, Captain General of the Royal Army, because of some affairs concerning the Count de Dona, and I went to an inn at Allieres, a few hours from Lyons. It so happened that our host and hostess, both healthy and full of vigor, had each in fact lived for 104 years.

In the previous discourse I have sufficiently shown by examples from the ecclesiastical and temporal history that there have at all times of the world been persons who lived for several hundred years. From this is it very easy to conclude that it is not impossible for us to live just a s long as they did, and that the story of Friedericus Gualdus, who was 400 years old, is no fiction. About this, I will now also bring here a literal extract from the Dutch newspaper of April 3, 1687.

Extract from a Letter from Venice of March 7, 1688: "Three months ago, a man by the name of Frederico Gualdo left here, who is 400 years old. He carried with him his portrait which had been made by Titian, already dead for 130 years! From this we may conclude that this man has had the true Universal Medicine, whereby he was able to keep himself healthy for such a long time. This is no fiction, however, but there are here many credible witnesses who spoke with this man himself, and who left here for no other reason that that it was said of him that he possessed the true secret of the Universal Medicine. The lovers of great curiosities will investigate this matter and inform us of the result, so that we can thereby also be useful to the public."

For my part I say that this Friederico Gualdo could reach such a great age wither without any medicine and solely through a well-regulated life and moderate physical exercise, and perspiration, or also through the Universal Medicine. Among all sayings, this Latin one is one of the truest: Plures gula occidit quam gladius. Reveling and immoderate eating and drinking strangle more people than the sword. Therefore I first begin with prescribing a rule of life for those who wish to live long and stay healthy. But I also want them to have a healthy body, and in the healthy body there should dwell a healthy or rational soul and spirit. They have to be of cheerful disposition and not subject to excessive emotions. Now follows what a person with the above characteristics has to observe.

First, he must guard against eating variously oddly cooked hard dishes and imbibing various drinks. Then he must well chew his food with his teeth, as that is the first digestive preparation which is effected by the moisture from the saliva of the glands of the jaw. When man is at table, he must eat the food and fruit alternately, so that moist and dry, fat and lean, sour after sweet, and cold things after warm ones, and vice versa are mixed together. For in this way the excess of the quality of one kind of food is always corrected by its contrary.

When a man has had a good drink or eaten fruit, he should eat dry bread afterwards. To counteract the too much wine he has taken, he can eat something acid or drink some lemon juice, which immediately gets rid of belching. The latter generally occurs after a meal, either due to all too much eating or even to great emptiness. Should he still be overheated from the wine, he can use cooling things but never additional heating ones, which could easily cause a high temperature. The strongly distilled spirits serve only to strengthen the stomach and promote digestion when one has eaten a bit too much; but if one gets hot from too much drinking, they are very harmful and dangerous. Aside from that, those distilled spirits are an excellent external and topical remedy. However, it should also be noted that after the use o this kind of beverage had been introduced into America, those peoples did and are doing quite a bit toward the shortening of their long life, just like ourselves.

As far as exercising the body is concerned, a man must never do violence to himself, unless necessity requires it, yet always according to the maxi: Ad ruborem, non as sudorem. Until you get warm, but not so much that you sweat, so that only the natural warmth is roused and the pores are opened, enabling Nature to cleanse herself through perspiration.

When a man has become quite hot and is tormented by thirst, he must not go to any cold place. He must stay put without making any movements, must bare his stomach or throw off his wig, but have a drink. He should drink wine rather than water, which could cause pleuralgia due to its coldness. But if he happens to be I the fields where he cannot have any wine, and he can no longer bear his thirst, he must drink by drops and not swallow in big gulps.

When getting out of bed, one must not immediately stand at or under the window, or go out into the cold air, because every quick change is dangerous: Juxta Hippocrates: Omnis repentina mutatio est periculosa.

When in a cold winter the nose, hands and feet are like frozen, one must be very careful not to put them quickly near the fire or stove, or dip them into warm water, because the members cannot without danger tolerate a change from one extremity of quality to another. It would be better to go into a heated room or a stable and thus get gradually warm by good movements, while simultaneously calling the escaped natural warmth back by means of the external tempered one. During my travels I used to moisten my socks of common linen in brandy every morning and took care to keep my boots wide enough to allow the blood in my feet to circulate freely right into my toes, which I always frequently moved. Aside from this, it is also customary to rub completely numb member with snow or ice-cold water and to move them well. Thus they will gradually recover without any danger.

When the time of the new fruit arrives, you must first eat but little of it, to allow the stomach to get used to t by and by and you do not have to worry that the intake of too much new fruit will cause fermentation of the gastric juice, resulting in many kinds of fevers. However, I do not wish to deny that there are some kinds of fruit which can be eaten without hesitation, as they do not produce much juice that hinders nutrition, and pass through easily.

Likewise, it is often said that a dangerous illness generally follows a quick change in the manner of living, to which travelers are especially subject.

Whoever does not share the views of Pythagoras regarding drinking, should resort to wine for a healthy and long life. Among different kinds of

water the best is that which is light and totally
without smell or taste. It is Emperor Nero whom we
have to thank for this discovery. Drinking water
that has been purified by distillation and then made
fresh again with ice immediately kills any nests of
worms produced in the stomach. And with such water
the learned Perreault, a member of the Royal
Academy, cured a nun as if by a miracle.

Insomuch as sleep is a necessary part of life and a
model of death, it should be sweet and calm. You
must not only slumber but also relax your thoughts,
as Apollonius of Tyana said to Phraotes, King in
India. That, however, is impossible for those who
drink a lot of wine, as its' rising vapors always
move and change the species of the imagination. That
is why those sleepers are completely tired on
awakening. Instead, those who drink much water (I
speak from personal experience) sleep much better
and sweeter, and their thoughts are so calm that
they also comprehend all things in their true nature
and form during sleep. Their sleep is neither light
nor heavy, nor disturbed by bad dreams.

Therefore Apollonius says in Chapter 2 of his life
by Philostratus that the idolatrous priests of the
Amphiaraus ordered those who came to their temple in
Athens tormented by dreams, to abstain from wine for

three days, whereby their dreams became much purer and clearer toward morning, so that they imagined having received them through divine inspiration and demanded an interpretation of them.

Irrespective of this, it is certain that if a man takes a glass of wine before resting and puts his head down, the wine may well move the species of his imagination by its vapors, while his mind is ever diverted from strong concentration on one thing, and in that way sleep is induced. This is well proven by the example of the famous Minister of State Tellier, Chancellor of France. In 1660, when he was overheated from the spiced Lenten fare and exhausted from much business, he used this means to get some sleep during his frequent travels to and from Avignon on business.

The physicians, in their Ars Longa, their Long Art --- which often makes for a short life --- stress three points in particular, which they call Diagnostic, Prognostic, and Cure. By means of the first they try to recognize the origin, cause and seat of the illness. Through the Prognostic and Cure they often prescribe remedies which produce a totally contrary effect to their intention. However, we cannot do otherwise, we must use physicians and their remedies, as the Scripture says: Honor the medical doctor because he helps you in need. But here we must use the shorter way of retaining our life for a long time, against the Long Art of Hippocrates, for whoever controls the beginning has won everything, as the poet says:

Principiis obsta, sero Medicina paratur,
Dum mala per longes invaluere moras.

Regarding the Diagnostic, we have to pay careful attention to anything new or unusual we feel, but it in the midst of our rest, or when we are about to rest, or when we are going about our business.

In addition, we must also carefully observe if any change occurs in the evening after work or when getting up in the morning. In the latter case there is more cause for worry as there is generally more strength and health after sleep and rest, through which our physical forces are renewed. In do doing, we must note is there is heaviness and tiredness of the members and if the appetite has been completely lost.

When now we feel weak and exhausted after ordinary work, I say that the numbness and heaviness of the members is due to all too much gastric juice, which is found in the viscera of all members and cannot be eliminated better than through perspiration and sweat. It can be produced by assisting the inner natural warmth with the external, and it is certain that, done in time, it prevents many an illness.

The art of sweating, however, consists in lying on your back, well tucked in between two white linen sheets and two eiderdowns, filly covered up, with only your face showing uncovered outside. You must not come out of this warmth until you have profusely perspired. After sweating, you must not get out of bed immediately but only half an hour afterwards. If you continue doing this for several consecutive days, your appetite and strength will return gradually, causing you to feel fine, fit and light in all members. For through this sweating the viscera are cleansed of all superfluousness, without pain or detriment to Nature. This cannot be done by any medicine unless it be the Universal medicine, which we will soon teach.

Therefore, for our health we can undertake this sweating three times a year, in the spring, fall and winter. When there is a lack of appetite in the spring or fall, we must eat little when sitting at table and do more vigorous physical exercise. But if the appetite has been completely lost, so that the mere sight of food causes aversion, we must not eat

anything for 24 hours, take a long walk, and help
the natural warmth in this way.

We must eat only small amounts of food that contains
much gastric juice or which is very nutritious, as
an excess of gastric juice generally causes fevers
and in children, the falling sickness, epilepsy.
Only those are cured who get rid of the excess of
this gastric juice with frequent vomiting.

Fever causes strong fermentation coupled with an
increase of the blood in both young and old, even if
they are accustomed to leading a dietetic and
regulated life. For as soon as they eat a bit too
much, their head becomes heavy, which is often a
sign and forerunner of a stroke. The reason is that
the fast and strong increase of the gastric juice
causes the blood to be frequently and forcefully
pushed to the brain where it tears the small and
subtle veins, thereafter spreads within the brain
and presses the nerves together. This hinders the
circulation of the vital spirits where, as Ferrelius
says, the natural warmth dwells. Interference with
their circulation can cause death if help does not
come in due time through the opening of a vein or
through sweating. Consequently, through the former
the cause ir removed but through the latter the
cerebral veins are softened, to enable them to

gradually expand and stretch again without danger of tearing.

As soon as we notice an indisposition, we must not lose any time and immediately use the remedies, thus preventing a long illness which might otherwise ensue. At the same time, we must think of the air we breath and the food we eat, and according to their quality and in consideration of our natural warmth decide which air and which food are most suitable to us, and in what way we can best help the natural warmth and drive the cause of the illness out of the inmost parts of our body.

For the sake of this cause, calm and warm air is to be chosen, that does not carry any repulsive smell. Places that are all too much subject to the winds are often not healthy, although all too warm places and those where the air is full of bad vapors require the wind to cleanse and refresh them, As is usually said about the city of Avignon: When in Avignon is not constantly windy, Morbona (illness) reigns there with its servants.

I have shown in my treatise entitle L'Homme Artificiel ou le Prophete Physique (The Artificial Man or the Natural Prophet), how very necessary it is for our health to understand the condition of the wind and the air, as in the view of Vitruyi, in Chapter 6 of his first book on architecture, the noon winds can cause incurable diseases, such as coughs, tuberculosis, diseases of the nerves and members, must as the residents of the town of Meteline, on the island of the same name, are best refreshed by the midnight winds.

In those who are wounded it is easy to observe that the pain of the wounds flares up and they become more sensitive with strong winds, because the external air causes the inner wounds between the skin and the flesh to be felt and hurt, due to its penetrating circulation.

When the illness is of long duration, the patient must be carried from his room into another. Then the windows of the previous room must be opened, fresh water must be sprayed about, after which the patient can be brought there. For the same reason careful Nature also frequently gives the patient the desire to change the air and location, which should not be denied.

As the patient gradually recovers, he generally has a desire for eating something acid, which is then very healthy for him, provided it is done sparingly and carefully, because Nature arouses an appetite for that which is right for her. This is so true that very often patients find their recovery through a moderate use of that which the medical doctor forbade so earnestly.

To say it briefly, diet and sweating are a part of the Universal Medicine, for Nature should in everything be our teacher from whom we must learn the right and true means for living healthy and long.

Because in childhood there is much heat, it throws off, by means of the smallpox and the so-called red needle-rash, whatever corrupted gastric juice and other impure matter there is in the body. When then that natural warmth has been decreased, we must stimulate it again, so that it may accomplish its effect just as it does in older persons.

Dancing, ball games, hunting, military, and other exercises arouse the natural warmth to do it share and drive out everything superfluous from the body through perspiration.

That is also the reason why farmers keep their health and live long is because they are continually working. As they do not make any debaucheries As they do not make any debaucheries, they do not know anything of podagra; which causes Seneca to say in Chapter 1, of his Hyppolitus: Only in rich Houses does podagra like to dwell.

For podagra in general avoids women and poor people and instead makes itself at home in palaces and rich houses where many good morsels are enjoyed. Therefore it has always been true that no hard working individual has ever had occasion to complain like Herod, who says: When I should eat, have no hands. When I should walk, I have no feet, but when I am supposed to have pain, I have hands and feet.

Just as every quick change is dangerous, lean persons who are putting on fat have serious grounds for preventing podagra through sweating.

Consumption, colic, and dropsy are cured with sweating. Whoever contracts the plague, and Nature has already pushed out some plague glands, can infallibly be freed from it through sweating. I can say precisely the same of leprosy, for if the natural warmth is sufficiently helped, it will completely throw out all superfluous and impure matters. Therefore those who have smallpox or the red-needle rash are rightly kept warm, and the natural warmth is aided with a confection of hyacinth, alkermes, or theriac (treacle). Gout itself, even if one no longer speaks about it, is cured by frequent sweating. Likewise the stroke, provided one starts early with bloodletting.

Just the same must be done for the trembling and shaking of the cerebral veins have ruptured. The blood has run out of them and it presses on the beginning of the nerves, thus granting the vital spirits their influence only intermittently. The only remedy is a good diet and sweating, as the serious blood that has run out thereby consumes itself. I have seen how some members suffering from a painful discharge were cured by putting them for a few hours in the hottest summer sun.

The best food for a sick person is good meat that can easily be digested. In fact, when it is cut into small pieces and the soft bones are minced quite finely together with the marrow, then pounded in a marble mortar, you must add to it something that has been found advisable, that keeps the patient's body open and is pleasant to his taste, sour, sweet, or the like, according to his appetite, then cook it over a moderate fire. After that, everything is strained through a cloth and you will have, as it may be called, the quintessence of what a patient is allowed to eat, also his beverage, according to whether he wanted it thick or liquid.

I am referring to those who wish to keep the volatile salt of the meat in the broth to the book De la Machine de Monsieur Papi pour Amolin les Os. There they will find my report, which I added in order to achieve the thing more easily. It has been printed in Paris by Michaler. Now I will also speak of the easy and sure way of preparing the Universal Medicine.

As we learn from the preacher Solomon that all recovery comes from God and that He causes the medicine to grow out of the earth (Altissimus creavit de terra Medicina), it is not necessary to ask here from whom this medicine has come down to us. Little do we care whether it has derived from the Hebrew Cabala or from the famous Dr Apollonius, or from his son Aesclepius, or from Hermes Trismegistus, from Raymond Lully, Arnbold de Villanova, from Bacon, from the Cosmopolitan, or from a Brother of the Rosicrucians. Suffice it to say that its composition is easy and of small cost, and that its effect is certain and assured, so that we could believe that it has the power to rejuvenate. This would appear to be irrational, if we had no certain an absolutely true evidence of rejuvenation both in the Holy Scripture and in secular history. The Royal Prophet, in Psalm 1-2: 5,

attests to the certainty of the matter in two
sentences. First, that the eagle becomes young
again, and secondly, that just as the eagle's our
youth can be renewed and brought back: Renovabitur
ut aquila juventus tue.

All Church Fathers believe that the eagle becomes
young again, but they are not of the same opinion
concerning the way in which it is done. In his
interpretation of this Psalm, St Augustine says that
when the eagle becomes old, the upper part of its
beak becomes so crooked that it can eat with no food
with it, or very little. Therefore it would crooked
that it can eat no food with it, or very little.
Therefore it would lose its strength and become weak
due to dieting all too long. But if it were to whet
off its said crooked beak on a stone and shed it, it
could again eat sufficiently, getting new strength
as if it had been completely rejuvenated. The
Prophet Isaiah speaks of this rejuvenation of the
eagle in Chapter 40:31, and Hiob in Chapter 39:25.
mention of it is also made by Aldrovandus in his
Ornithologia, Book I, and Gesnetus in his 5th Book
on Birds.

Likewise, it is also known that snakes slough off
their old skins, which are sometimes found in the
bushes. I will here not speak much of the big

grasshoppers and their sloughing. I myself have encountered it in Dauphine when I traveled there to have a look at the most beautiful and highest bridge in the world. It has only one arch and reaches from one mountain to another, from where the wind appears at certain hours from an unfathomable hole in the rock, blowing along the river as far as Orange. Thus we also read in Philostratus, Book 3, Chap. 1 of his life of Apollonius of Tyana that there exist on the highest cliffs of Mount Caucasus, which no man can climb, a special kind of monkey which gather the pepper for the inhabitants. The meat of these monkeys is an absolutely good remedy for old and sick lions, for if they devour one of these animals, they recover and become young again.

If then birds and animals can become young again, we can reasonably infer that it can also be possible for man. When we are born, our constitution is quite warm and moist, while in old age it is cold and dry. It is therefore only necessary to replace the radical moisture and the all too great dryness of old people with the moist constitution of youth.

Now I must really show that individuals have indeed lived who became young again. Medea, well skilled in medicine, rejuvenated the old Jason. That is why Ovid said in Book 7 of his Metamorphoses, that Medea had Jason chopped and boiled, which is to be understood as referring to baths prepared by her with minerals and other ingredients. And this is not contrary to truth, because Petrus Martyr Augerius, a ma from Milan, asserts in his Decadibus that there is a well on the island of Bonique which causes old people drinking it to become young and strong again, although their hair stays grey and the wrinkles in their faces do not disappear. There is supposed to be a similar well in Lucayam according to a report by Petrus Chieza (Chapter 14, Part 2, Historie de Peru). We can also look up what Herodotus writes in Book 4 about the power and virtue of such waters, which have therefore been called fountains of youth.

In the first discourse of his Horti Floridi, Torquemada reports that in Italy, in the year 1531, an old man of 100 years who, as they say, already had one foot in the grave, was once rejuvenated in every respect and lived another 50 years thereafter. He says the name of another old man, which the town councilor of the place himself attested to. Valescus Tarentasius writes that he saw in Monvedro, also called Sagonza, in the Spanish kingdom of Valencia, an abbess who creaked with old age, had no tooth in her mouth, and was walking around deeply bent. Then, he writes, her teeth were growing again, her hair was becoming black, the wrinkles in her face disappeared, and she got a beautiful smooth forehead like a girl of 15. Yes, in a word, she had become young again.

Two more recent and credible historians, Ferdinand Castana in his 8th and Petrus Maffeus in his 11th Book, testify that a noble red Indian had rejuvenated himself three times within 340 years, the time he had been living. This report is sufficiently authentic, as Mendoza assures us in his Viridario, 4th Book, 17th Problem, that various Jesuits had known and spoken with this three times rejuvenated Indian, and also confirmed it in their letters.

Now we will speak of the Universal Medicine and its composition, after first declaring that it does not consist in any alkali or acid, which are two principles only recently talked about.

If we are to believe Takenius and after him his new Hippocratic-Chymical sect, it is possible to become a great medical doctor all at once and without much studying and pondering, acquiring very fast a great reputation. For nothing is required except a man have a good knowledge of acids, alkalis, and opiates. When the patient is hot as if he were lying in fire, he must be given an alkali to stop the fire. On the contrary, an acid must be given to someone down with frost and cold in order t arouse the natural warmth. Opium, however, is to be given to the patient to provide him with sleep and rest when, for instance, the pain is great. In fact, many attain to glory and a great name through the quick help and relief they are giving the patients. This I have seen that one of them cured a catarrh with a great Universal Sweat, by grinding in a marble mortar a certain kind of turnip, called tenderette in Paris, and applying the stuff on the soles of the patient's feet.

However, the Universal medicine cannot consist of an alkali, or opiate. They only do away with the patient's symptoms but cannot remove the cause of the illness which is due to a humore peccante in the inmost of the members and must necessarily and above all be eliminated.

When the humores peccantes or malignancies, or other poisonous substances are volatile and subtle, they are to be driven out through the pores of the body by perspiration. But when they are mostly moist, through sweating. When they are moist but not subtle, they are eliminated through the urine; when they are less moist and on the contrary more material, the elimination can be done through purging or vomiting. The patient must not be weakened through purging, however, but his nature is to be strengthened.

Now I Have Come To Speak About The Required Qualities Of The Universal Medicine:

A Universal Remedy must have a kinship and likeness to our natural heat and radical moisture, partly to maintain them but partly also to replace them. Thus our exhausted forces are increased in such a way that Nature will without urging and of herself drive out of the inmost of the body all the things that are bad and contrary to her, be they acid or alkali, or coagulated, which causes stitches in the side (pleuralgia), catarrhs, podagra, etc. It generally happens that a man gets much overheated by strenuous exercise, resulting in cold through breathing air that is too cold and drinking liquid that is too cold. Afterwards, in cooling, the nitrous parts intermix with the fermenting blood in the lungs, and when there occurs a slight concentration of the blood, some of it is expelled at the end of the arteries.

Now and then the matter expelled attached itself to some part of body and causes pain in the nerves, due to the sharpness it had acquired. This continues till the natural warmth of the body dissipated or expels the sharp part. If this cold inhalation takes place at the time of the digestion, it can also cause an extravasatum through the lung, which mixes with the chyle and is conducted to a spot where it is not only so easy to dissipate it and is the cause of arthritic matter.

Accordingly, the Universal Medicine has to evacuate everything alien to Nature, everything in the inmost of the members, through perspiration, sweating, or urinating, but seldom through defecating, and still more rarely through vomiting. Ordinary medicines have no such effect. They attack with overheating and with poisonous sameness do violence to the sick matter of Nature, forcing her to get rid of her enemies, almost against her powers. Furthermore, the Universal Medicine must be composed in such a way that it can be administered at all times, to every constitution and every age, to both children and old people, without its more or less strong dose being harmful. It is supposed to cure the most desperate illness after taking it a few times. It is also supposed to be a general remedy for all external damages. There now follows the preparation of this Universal Medicine.

Composition of the Universal Medicine

Take some purified Air-Salt, put it into an iron vessel and let it gradually melt. When it is melted, pour on it some finely ground lindenwood charcoal which will immediately burn and be consumed. That is why one has to let it burn gradually till the salt is almost fixed after the detonation, and it gets a somewhat bluish-green color. This happens when the charcoal no longer puffs up as it did before. There after, pour the melted salt into a marble mortar while it still quite warm. When it has cooled, it is as white as alabaster and brittle as glass.

Now pound it small and sprinkle the powder on a glass slab or table but cover it to prevent any dust from falling on it. Hand it in the air, at a place where neither the sun nor the rain nor the white frost can touch it. Put a glass vessel underneath it, into which the oil that will drip fro it can fall, because the humidity in the air resolves this alkali within 14 days.

You will find that the oil is twice as heavy as the salt previously was, provided this process is undertaken at the right time, that is, when the weather is neither too cold nor too warm but moderately humid. Then the salt attracts the air invisibly, just as we draw the breath into us. If this oil is well rectified according to the art, it is an excellent and powerful menstruum for extracting the essence from various mixtures.

Now take 4 or 5 parts of this rectified oil and one part of the best antimony. It can be recognized by a certain redness which it has from the gold, to whose ore it is close. Pound the antimony on a marble to a very fine powder, put it into a cucurbit and pour your rectified oil on it, in such a way that two-thirds of the glass stay empty. Quickly close the glass so that no air gets into it, let it digest in gentle heat or at an oil hanging-lamp, until the oil floating on the antimony gets a golden or ruby color. Then pour the oil gently off, filter it through paper, put it into a phial, and pour on it an equal weight of the rectified spirit of wine. At least two thirds of the glass must be empty. Stopper it tightly, let it digest in small heat for several days, till the spirit of wine has attracted all the tincture of the oil.

Thus the oil will stay quite clear and white at the bottom, but above will float the spirit of wine with the golden tincture of the impregnated antimony. Separate from it the menstruum, by decanting, which can always serve for several processes and operations in order to extract the essence of antimony as often as desired.

Put your tinctured spirit of wine into a glass alembic, abstract and distill it gradually, till only about one-fifth that contains the tincture of antimony is left at the bottom. Or distill all the spirits of wine, so that nothing stays at the bottom except the essence or tincture of antimony. Thus you will have the Universal Medicine either as a powder or a liquid, with which you can cure all illnesses or preserve yourself from them.

When it is used as a liquid, 5 or 6 drops of it are taken with wine or bouillon, or in any other liquid suitable for the illness.

When it is used as a powder, 3, 4, or 5 grains more or less are taken. For even if the dose is a little stronger or weaker, it is not harmful like the medicines which all have poisonous qualities. The sick become healthy after taking it 2 or 3 times. But is the illness does not yield, the dose must each time be increased, and this has to be done three times a week.

This medicine cures all, even the oldest and most severe illnesses, such as the 4-day fever, dropsy, and the falling sickness (epilepsy). Yes, it not only cures all internal but also the external illnesses, such as wounds, ulcers, and cancer, when it is spread over them like a balm. It serves no less for deafness and eye troubles.

Finally, this medicine drives all headaches away and helps digestion. It is a true Aurum Potabile (potable gold). It generally works quite invisibly through the perspiration; more frequently through vomiting. Consequently, its effect is quite natural and without force so that the patient does not become weak, exhausted, or debilitated, as usually happens with other medicines. It can therefore be used at any age, at any time, and irrespective of the constitution. Use it and make it known, but especially let the poor benefit from it, and praise God Almighty Who has created the medicine.

The Revelation of the True Chymical Wisdom
J. J. Chymiphilo

That is the Accurate and Sincere Disclosure of the
Materia that has to be Taken if One Intends to Make
the True Philosophers' Stone, Lapis Philosophorum,
the Universal Tincture.

From many Theophrastic manuscripts, preciously never
printed, shown in the plainest and clearest words,
so that it could not be clearer, Also for the sake
of those who are not yet much experienced in the
Chymical Manipulations, provided with special
instruction and the shortest possible directives for
the necessary preparations.

Everything openly published for the pleasure of the
Lovers of Chymical Wisdom.

By J.J. Chymiphilo,
Anno 1720

Those who know the Art of making gold keep it as secret as possible, though they reveal it sometimes in a strange way, as is shown by the following story.

A special booklet has been published, dealing with the infallible knowledge of Friedericus Gualdus to make gold. The gist and an extract of that booklet are here presented. The author writes as follows:

I was still a young boy, when I met Herr Friederich Gualdus for the first time. However, my young age did not allow me to recognize his enlightened intelligence as well as his profound erudition. Neither do I know if he had then already been a long or a short time in our town, but I only remember his face and whole being represented a man of about 40 years. He kept that figure all the time, without change, till the year 1680, at what time I was obliged to leave for Naples on account of my business.

At first, he behaved like any traveler and was either completely alone or had a small boy in his service. He was living in two average rooms which were not fully furnished and hardly comfortable enough for a good student to live in. Nevertheless, he made friends, I do not know how, with some noble gentlemen of this State, who, although they owned various mines, could not continuing exploiting them due to the losses they had suffered. He began to help with such a considerable advance of money that they soon disposed of 60,000 ducats.

In the meantime, he fell in love with a young girl of this noble house, although she was still a child. Her mother, who was very grateful to him because of the benefits they had received, through which her mine and at the same time her family were again on the road to prosperity, also felt that she could not prove her gratitude better than if she became related by blood to such a great benefactor. Although the child was not yet fit for matrimony, she nevertheless set a later date for the marriage, thus getting enough time to draw up the marriage contract in proper documents, and she promised a dowry of 16,000 ducats. To show his consent, Gualdus in turn agreed to return the same amount (if the marriage did not take place). He also meant to bestow on his fiancée still much greater riches than he had already granted her parents.

Just as it is only too true that noble gentlemen often change their minds, so the said aristocrats, after attaining a higher status, to which the benefits of Gualdus had especially helped them, considered themselves far too good to make friends with an unknown man, and immediately tried to undo the deal. Such an unfair procedure hurt Gualdus very much. He withdrew from their friendship and demanded that the money he had laid out should be returned to him. As many difficulties were put in his way, he tried to make legal steps to recover his property. However, to cut off every further annoyance and quarrel, he consented to relinquish half of his claim in a settlement giving him instead a beautiful and honest testimony.

So that those noblemen should learn that they had lost a great deal with his friendship, he took very uncommon steps to be raised into the Venetian nobility. With this in mind, he proposed to the most high Republic to undertake a great, useful, and profitable work at his expense, if he were in return raised into nobility, which many acquired at that time by paying the sum of 100,000 ducats. But there existed quite irremovable obstacle to this proposal, insofar as it was not in accord with the sovereignty of the Republic to give a reward to anyone before he had earned it. Therefore Gualdus decided to pay those 100,000 ducats, but in the name of a depositor and on condition that they be restituted to him after he had carried out his abovementioned

proposal. Otherwise the money would be left to the
Republic.

In the meantime, however, he was raised into the
Venetian nobility under the pretext of some merits,
and declared a Patrician, and not like the others
who had paid a great amount of money.

But the Council, which kept very strictly to its
wise rules, did not wish to introduce an innovation
with that admission and again cancelled the treaty
it had made with Gualdus, although his name had
never appeared in it, and in such cases only the
words "an unknown person proposes" are used, and the
man is only named when the proposals are accepted
and the conditions fulfilled. It was nevertheless
learned subsequently that Gualdus had proposed such
a work and had also accomplished it.

After he had dedicated his mind to wisdom and
withdrawn his thought from vanity and love, he
sought his peace wholly and solely in the true
wisdom, and abandoned himself totally to it.

Only after this, his deep knowledge and high
intelligence were recognized in the societies of
learned men where he was present several times. For
when they were discussing philosophical matters, he

always knew how to dissolve the syllogism effortlessly and to put the disputing parties to shame. When they spoke of political and world affairs, there was no one with a better knowledge of the most secret government intrigues, or better acquainted with the cabinets and secret council chambers of high potentates. Was theology at stake or common law, he proved that he thoroughly knew the most hidden canons and most famous verdicts. Aside from all this, he was no less an accomplished astronomer and also uncommonly experienced mathematician. In short, no science could be discussed that he did not thoroughly understand. Yes, the stories of the oldest times were as fresh in his memory as if they had only happened today. Thus he also spoke various languages perfectly, that is, Greek, Hebrew, Latin, French, Italian, and others aside from German which was (he said) his mother tongue.

Such extraordinary qualities could not stay hidden but come to the attention of many learned men and persons of rank in Italy. Among them, some had already heard a great deal about him when they arrived in Venice. Aside from trying to see rare things, their special purpose was to see Mr Gualdus and to do their utmost to obtain his friendship, which they subsequently kept up with letters. Yes, some only went to Venice so as to continually associate with him and have the honor of being called his scholars. This all the more after the rumor was circulated that he possessed the Hermetic Treasure, owing to the fact that his portrait had been seen by many painting experts who asserted that it had been done by Titian's famous hand. Already a long time ago he had markedly decorated his residence with fine paintings, putting the most recent in the best rooms. Behind the door, however, he had hung the said portrait. One day some people, among whom there was an experienced painter, came into his house to look at his pictures. When the painter saw this painting behind the door, which had just been closed accidentally, he was amazed and called out loud, "This is the hand of Titian!". Gualdus laughed and said that if it were true, he would be over 200 years old, while he was only 86. This happened in the year 1677. The painter, however, would not stop but continued asserting that it was the hand of Titian, although Gualdus pretended to be 86 years old and although he only looked like a man of 40 and was also able to do all those things which a man can do in the prime of life.

This rumor was the cause why people became firmly convinced that Gualdus possessed the great secret.

Indeed, at this time Mr. Marques Santinelli applied for the first time for his intimate friendship, seeking it in all possible ways, and finally he obtained it. Petrus Andreas Andreiny, a Florentine nobleman, did no less. In Naples he was famous both for his wealth and for chiefly collecting old medals, coins and other curiosities. Therefore, when shortly thereafter Mr Marquese had a small book printed at his cost under the title of Androgenes Hermeticus, etc., it was said that it contained Gualdus' teachings, since few books had been written so well and emphatically about this science as just the said small work. However, Mr Marquese must in this not be denied his share of fame, as his incomparable sonnets printed in the same booklet are no mean ornament and can easily make us believe that Androgenes itself was his learned offspring.

Aside from those, there were various other very learned clergymen who resorted to this Oracle with their questions. Because I knew one of them, who was both of noble extraction and belonged to a famous Order, and his letters and ensuing answers passed through my hands I do not consider it wrong to add these here, especially as I am sure that this will please all learned and intellectually curious persons.

assurance that if Good should bestow His Grace upon me, I would immediately express my due thanks at your feet and subject my will to you in everything.

As I have now learned, the whole difficulty of our Art consists in the preparation of our water. I have seen the Sun and the Moon in the first philosophical sublimation of our water and have drawn it off 7 times from the moon. I would like to know whether our Mercury, after being purified 7 times, is then perfect and capable of penetrating through the pores and airholes of the Egg, and of dissolving the Sun. Or is another manipulation still required to turn it into powder, such as the relocation? My doubts are caused by the fact that it is not snow-white and there is still some of its smell about it. As it is not fixed when it ceases to smoke, how then can it make things fixed? Therefore I pray you for God's sake to tell me if the thus 7 times purified Mercury is perfect. If it is not, what else is required? The other difficulty that I encounter lies in the feces of our sea, from which in my opinion I have extracted the saltpeter which, however, is so sharp that the nose cannot withstand its smell, I do not know therefore, for what I should use it, seeing that the Mercury is purified.

Our cinnabar does not require any ferment, as Nature has already made it perfect, though it becomes much purer when purified through the Art. Of what use then is salt? I believe that it must serve to ferment the water. Would you therefore please inform me if all the water is to be fermented, or only that part where the snake penetrates. Also, what weights and measures are to be used to do it? For meanwhile I am thinking that one part of salt and ten parts of Mercury must be taken, although I am in this case not aware of the proportion of the weight. Aside from this, I would also like to know how much of the Mercury must be prepared, so that the imprisoned child does not lack milk.

I have come across another difficulty on account of the form and shape of the vessel. But I am speaking of the last fixation, when I believe that the form must be like a chicken's egg, fully filled and sunk in so much that the snake's head does not rise above it. But while the philosophers say that it must have a long neck, I would like to hear about this from your mouth and be informed whether part of the snake must show out of the menstruum and the neck stay inside. For this causes me some thinking, as I am afraid that the air of the neck would hinder the generation or birth.

As I usually take into account all contingencies, I would also like to know if it is better to take the gold and silver direct from the mountain, to enable the sympathy to manifest itself all the more strongly, or to take the gold only from the mountain, and the Moon from the hill.

Furthermore, it is stated for certain in the books that the Phoenix, when it just comes out of its nest, must be captured thus pure and that not the least little thing must be added to it or taken from it. But if there should be a mistake in this, do deem me worthy of being reminded of it, and if you feel that God is pouring His grace over me, kindly help me and also reply to my above-indicated doubting questions without naming the authors. But neither despise my stupidity which causes me to write this with so little purity, but which is nevertheless a gift of God who is well aware that neither profit nor another worldly intention but a mere desire to know something and subsequently to apply all the knowledge to the honor of the Giver thereof, has impelled me to this study. Do therefore with me as God will inspire you, as I am assured that He who has inspired me to write to you has also given you the kindness to satisfy me. Finally, I remain with my whole heart, etc., Sir, Your most humble and obliged servant, D.C. von R.

Please inform me also how the vessel can be cared for so as to last for 9 months.

Naples, 28 July 1678.

Reply of Mr Friederich Gualdus to the Preceding Letter

Sir,

Your letter has been handed to me, and although you are not known to me, I have sufficiently recognized your great intelligence from your lines. Meanwhile I much regret that I am not qualified, much less able, to answer your serious questions. In this I am the more excused as I am well acquainted with the Italian language. But to satisfy you nevertheless somewhat according to my little understanding I first say that you are speaking all too obscurely in your letter, like the true philosophers, and you therefore mix everything up so covertly that one does not know how to reply. There is a good deal to be said for the fact that on the Damascene Field that Virgin Earth is found which is the material of our magistery and work of art. And I will add that the said Virgin Earth has never seen either the sun or the moon, although it contains the Sun and the Moon. However, you do not say how this Damascene Field, let alone this Virgin Earth, as the material of our work of art, is constituted. And how and in what manner it can be obtained.

Furthermore, you write: Thus, our living gold is well known to me, by what it is dissolved, as well as the sympathy that develops between both, because both descend from one father. All this may very well be so, but you do not report of what kind this living gold is and this water that dissolves it, s I cannot give you a considered opinion on it. In the same way you are also confused about I do not know what difficulties, which is due to the fact that you do not work with the right material. In addition, you mention that the whole secret consists in the preparation of our water, and that the Sun and the Moon are contained in it, which is all pure truth.

Regarding the difficulty which you encounter with the sevenfold sublimation of our water, that is, whether or not it is then perfect, I reply as follows: The philosophers want us to distill the water 7 times, but they say septies, aut pluries (seven times of more). And Sendivogius indicates when it is perfect, namely, when it leaves white feces at the bottom, so that is the unmistakable sign of its perfection.

You find the second difficulty in the feces of our sea, from which you have extracted the salt. From this I feel that you are so much mistaken in the material as in your manner of working, thus creating many difficulties. Therefore, if you had worked with the right materia is at hand, the right vessel follows afterwards of its own.

But as far as your other questions are concerned, namely, whether it is better to take the gold and silver from the mountain or solely the sun from the mountain and the Moon from the hill, I do not know what peculiar talk this is. I believe that you are confused in everything, and I cannot understand whether your words are puzzles, figurative speech, or fiction. Therefore I do not know how to reply to them. Pardon me if I speak frankly and without hypocrisy. In the meantime, if I can help you with something else, I am ready to do so upon a hint from you. I remain, etc., Sir, your most devoted friend, Friederich Gualdus.

Venice, 1 September 1678.

Second Letter of D.C. von R to Mr. Gualdus

Sir,

I thank you very much that you found me worthy and replied to my ignorance, which is really an effect of your generosity and not of my merits. And I consider your letter no less than an answer given by the Oracle, although you have not replied anything essential to my questions concerning the Work, and I can see from your polite lines that you have not clearly indicated how the substance of the materia may be well known to me. But after obtaining the knowledge of that materia from the Lord of Truth, I only wished to get some information on how it has to be treated and approached, which someone who does not ask for anything else can easily obtain.

Therefore, in order to obtain it, I said that I had found the Virgin Earth in the Damascene Field. Although are saying exactly the same you then add that I had not named it. And although I am not allowed to name it (especially in a letter) because it has never been named by any philosopher, it is still sufficient that I said that I had found it with the help of the lamp of Diogenes. But if you nevertheless demand that I name it, I say that its real name is: Our water, our gold. It is that water which the philosopher rectifies seven times and the menstruum of our gold, that is, of that gold which is the Virgin Earth of the Damascene. As it has been formed by Nature, it is also put into the rectified water without any deduction or addition, to dissolve, sublimate, and calcine. This is the whole magistery or work of art, to fathom which God's Mercy bestowed His Grace on me after great expense and damage, for my incessant prayer and efforts.

Our material has various names, just as the place from which it it is names in my previous letter, after reporting that the water and the gold are always of the same kind. As I had spoken in a veiled manner, I wanted especially to know if I had to take both the gold and the water from a mountain, or the water only from a mountain. That difficulty or doubt arose from the fact that Morienus writes: That which contains all things in itself does not need any other aid, which is contrary to the opinion of other philosophers who take the water from the small mountain.

Sir, I have sufficiently disclosed my thoughts, Just then as our magistery can be compared to human generation, and yet, children are not always born, although I took refuge with you, as the Oracle, so that I might be instructed by you in the manipulations and would consequently not be mistaken if I wished to take up the Work. Seeing that I cannot find anything in the books that would satisfy me.

In particular, however, I would like to know what I am to do with the feces. While all pretend that the water must not be rectified more than seven times and the feces were left at the bottom. What then am Ito do with them? If I pour fresh water on them and distill them till they become white, I go against the rules of the most experienced philosophers and nevertheless doubt that they become white thereby. Yes, I also worry that the water might lose its power. But supposing that they become white, what am I to do with them, because they are highly esteemed by Hermes? Am I perhaps to put them back into the water, now quite white, while it ripens the warm gold? According to the saying of Brother Basil. As I am at it, must its earth be fertilized or not?

I pray you to come to the aid of my ignorance and to show me clearly, and not in riddles, where I may go wrong, and also give me your instruction in the greatest difficulty but especially concerning the vessel of the last fixation, as I do not know into what the gold must be put, if it has to have a long neck or be shaped like an egg; likewise, is it to be completely full or empty in the neck; also, if all the material has to be immersed or one-third allowed to protrude; and then if the vessel can last for 9 months.

I pray you, remember what Solomon said in his Book of Wisdom, which he imparted without begrudging, and be assured that this time you will gratify a man who is not altogether unworthy of such a favor. I am therefore expecting complete instructions from your kindness, how I have to operate, as also the removal of my other doubts. Under God's favorable protection commendation. I remain with all respect, Sir, Your obedient servant and student. D.C. von R.
Naples, 8 October 1678

Reply of Mr. Friederich Gualdus
to the Above Letter

Sir,

I am in receipt of your letter and have understood from it, no less than from your first communication, that you do not possess the right material, but have a vain and erroneous opinion of it, so that everything you undertake with it will be a useful and fruitless effort, Pardon me if I speak freely, I cannot be hypocritical but am telling the truth. If you were well acquainted with the right knowledge of the real philosophy, in which both the material and the operation are kept strictly secret, it would not be so difficult for you. True, it is not without reason that the philosophers are keeping every thing very secret, both the material and also the operation. But one depends entirely on the other, so that, if the material is known, it is also easy to know the operation can also easily be familiar with the material. There exists in the world only one material for which all operations taught by the right philosophers are suitable. Therefore they have not only kept the materia secret but necessarily also the operation. But, as mentioned, one depends on the other, which is the reason why it cannot be clearly said in words, far less described in letters.

This divine and sacred science is acquired in only two ways, although through divine revelation or through the instruction of a faithful friend. For to attain it by studying books is almost an impossibility. Therefore also, whoever wants to achieve it with many operations and various tests, will never reach his goal. For this science is just

like other sciences and can quite certainly be
learned without tests and manipulations, by
understanding it with the mind. This is so true that
it cannot be otherwise but must necessarily be so.
Thus one also knows before undertaking the operation
what kind of a thing it must be, and it is also
possible to recognize by certain signs if one
operates well or badly. Yes, one does not commit any
error or mistake, but everything will completely
agree with the rules of the best authors, so that,
whoever receives in this the divine revelation and
masters this science, cannot go wrong.

As I do not see from your letter that you know the
true materia, I cannot explain more about it or
speak about the operation. I only say that the
material is so wonderful and that whoever possesses
it simultaneously has in his possession the vessel,
the furnace, the fire, the menstruum, the gold, the
silver, the Philosophical Mercury, and everything
that belongs to the Philosophical Work. From this it
follows that your questions rest on vain sophistic
questions, and I will answer them in all liberty.
For today, nothing more except that I remain at all
times, Sir, Your most devoted servant, Friederich
Gualdus.

Venice, 2 November 1678

Another Letter of D.C. von R. to Mr. Gualdus

Sir,

Oh! That I might travel to Venice and discuss with
you! I would show you that my knowledge is not vain
or imagined but true and based on the right
philosophy. I would explain what great things the
mineral kingdom contains, and also, palpably
demonstrate how the radical moisture of the metals
is constituted. I would discuss about the great and
little world, and relate all the special features
found at the creation of Adam, nor omit the least
bit of the quality of the Damascene Field. I would
indicate what is the Philosophical mercury, what are
gold and silver. Yes, I would not leave untouched
any of the most famous philosophical riddles but
examine each most meticulously and show how clearly
the scholars have spoken of it. But as I do not have
the privilege of doing so, I must necessarily be
quiet, especially also as I do not dare to deal with
great things in letters. I am only telling myself
that where the ray of knowledge has once begun to
shine, it can impossibly be obscured by the clouds
of contradiction. Of this we have an all too cleat
example in Trevisan, who, having once recognized
truth could never be turned away from it by others
who tried to confuse him out of envy.

I know for sure that you would indeed understand that I am not wrong if you possessed such a great treasure. Do not believe that this is mere speculation, but it is the true realization ex visceribus cause (from my guts). Even if I have so far not attained perfection, it is not due to an error but solely to the fact that I have not yet undertaken the Work, although two years have already passed since I was illumined by God. Consequently, I do not know by what secret power I have been held, causing me to be satisfied only with science, not bothering much about the rest, and only preparing our water with the greatest pleasure and content. Therefore I have nothing to complain about and rather hope that God, qui dat esse et perficere, after letting me come to the cognition of truth through His Mercy, will also grant me His help in achieving the perfection of it. The reason why I enjoyed the acquaintance of such a great man as I esteemed you to be, was that I had kept back and deliberated upon the random questions I had presented to you but subsequently was bold enough to trouble you with them in my letters, knowing well that it pleased the most famous philosophers to find capable individuals as their students to impart to them that knowledge which is of no use to anyone in the other world.

For in just that way Morienus and others had been accepted and instructed by their teachers, However, it would have been stupid of me to try and obtain the knowledge of the true material, which had never been my intention. Instead, I tried to get an explanation about the vessel, the last fixation, and the feces of our sun. Do indeed not tell me that whoever knows one, must also know the other, as such a great difference between generation and science has occurred that many who have worked very much with the true materia, spoil and destroy it because they are treating it wrongly.

Although I have learned from the generation of metals how to imitate Nature in the sublimation of our water, I still require a yet better light if I am to reach perfection. I well know that our living gold is not always dissolved in its water, the reason being that the water is perhaps not always good. Therefore I only beg you to inform me how the vessel must be formed (I am speaking here of the last fixation), whether, that is, it should have a long neck or be in the shape of a chicken egg.

Nonetheless, I think it should have a long neck, although I do not like this because, if it were formed like a chicken egg, not the least bit of air would touch our gold which would be closed like a hen's egg. Also, like the latter, it contains the Mercury, Sulphur, and everything required for our magistery or Work of Art, and performs of its own all the operations described by the philosophers in so many ways. In its operation, we need do nothing except keep the heated water in its natural heat and pour some more on it when the child, or the Dragon within, begins to feed on it. In so doing, the same proportion must always be used.

Therefore, you can probably very well give a truthful answer to these questions without revealing our water, in which resides the whole difficulty; likewise, how the purified feces are again united with the water, which is quite different from the knowledge of the materia, so much so that one can very well be known without the other. And because I rely on your courtesy to receive at least one single recipe, I have not given credence to one or another babbler. I hope, therefore, not to oppose truth. And even if the Supreme in His just judgment should not deign to favor me with the accomplishment of the Work and in so doing use my services, I shall nevertheless die quite cheerfully after cognizing and seeing the truth, like the philosophers who are only badly screamed at by unwise fools for heaving spoken the truth.

Forgive my boldness of inconveniencing you with my ignorance, But I know for sure that you have understood what I have written to you, so that you do consider me quite such a great fool. No more for today, except that I remain, aside from commending you to God's protection, etc. Sir, Your most devoted servant, C.D. von R., etc.

Naples, 3 December 1787

From the aforesaid I now believe that everybody will understand that there was something uncommon and extraordinary about Gualdus. Here was a man who had lived like a poor student upon his arrival in Venice; who had never carried on any trade, thereby acquiring great riches; who had possessed neither goods nor revenues, and yet had loaned 60,000 ducats to some families, in addition to being able to advance 100,000 ducats for being raised into the Venetian aristocracy. He had sought the latter in quite a special way, as otherwise he would have had to register his arrival and age. Of the latter, however, his portrait, said most assuredly to have been made by Titian, is an undeniable testimony. After additionally spending many thousand ducats for the accomplishment of the work he had proposed, he finally also decorated his residence magnificently, while at the same timer performing other generous deeds. We are therefore not unjustified in believing that he must have possessed an inexhaustible treasure. As he knew simultaneously to keep himself in continual perfect health and at an invariably manly age, we must necessarily conclude that this treasure was that great medicine which extends its power over all three realms, the animal, vegetable, and mineral.

But the above-mentioned cleric did not possess this treasure, as his life ended a few months after his above-quoted letter, in which he shows such great intelligence and boasts as if he had obtained the right and true materia. However, if such had indeed been the case, he himself would still be here to speak in protection of life, and would have brought to perfection that great magistery or Work of Art for which he had so zealously striven.

In contradistinction, our Gualdus (or better, our hero) must without doubt be such a one. He gave the clearest signs of it when he finally departed from this town on 22 May 1682, perhaps for no other reason than that he had heard that his virtue had been revealed and made known everywhere. Before leaving, he had given full powers to his servant and instructed him in what he had to do with his assets. That same day, unexpectedly, he had some underwear and clothes packed into a small box, as if he intended to go for a trip to his country estate near Treviso, saying that he would be back in a few days. The servant to whom he had commended the whole house with the most precious furniture, had to stay behind, however, and he thus went away all alone, without company, or rather, he disappeared when he was 90 years old, as he had said of himself, or perhaps several hundred years.

The servants waited many days for the return of their master, but they did not see more of him, let alone receive letters, Finally they assumed that his journey had not been made to his country estate, as they also learned that he had not even arrived there. Therefore they disposed of part of his personal property according to the order he had left them, but they kept the rest for themselves. In so doing, they provided so well for their poverty that they were relieved of service ever after.

This then, is the whole report that can be communicated to the world about such an unusual event. Nevertheless, I wish to add here a few other letters which this great man wrote and of which I transmitted the original. By them we can recognize even better how certainly ad undoubtedly he was in command of the Hermetic Art. To others, however, who might have the good fortune of surviving him, I will leave the honor of describing his course of life more precisely, etc.

Letter of F. Gualdus to Mr N. N.

Sir,

You understand extremely well how water is to be reduced to or made into earth, of which one is dissolved and the other hardened into a shiny marble, while out of the latter the foliated Earth is sublimated. But the said reduction is done forward and above its own earth and with its own water, which has been drawn from it, however not with the lunar calx and the Mercury, which never combine together in such a way that they cannot again be separated from each other.

By body-calx we mean our bodies, which are alive, while the bodies of the mob are dead. They no longer eat and drink; the tyrant of the world has killed

them: Out of man grows man, and out of gold, gold;
yet not out of dead but out of living gold.

Our destroyed earth, forsaken by all spirits, is
silver, and is our living gold is again united with
its spirits, the shining earth will grow from it.

You have made the luteballs (limeballs, glueballs)
very well, and I hope that they will turn white.
Everything is also quite airtight with the other
earthenware vessels which become pale-yellow,
because the pale will always cause the yellow to
diminish, and they will always tend more toward
whiteness. It is easy to give the Stone but
exceedingly difficult to make it. It has to be
obtained with pains and toil. Then it will be
considered as that which it is.

I remain with all my heart, Sir, Your very devoted
most willing friend, Friederich Gualdus
Venice, 11 September 1677

Another Letter of F. Gualdus to Mr N. N.

Sir,

From your letter I see that things also go well with
the boiled Mercury, which is so and not otherwise,
nor can it be. I will tell you the reason why.

Mercury cannot be hardened or turned into earth
except by a certain degree of the fire especially
required for it, which is called the fire of
Mercury. But what kind of a thing is this fire? It
is not, and cannot be, a mild fire, for even if
Mercury were to stand for a thousand years in a mild
fire, for even if Mercury were to stand for a
thousand years in a mild fire, it would not harden.
Neither is it, nor can it be, a strong fire, for if
one were to put Mercury turned into earth into a
strong fire, it would become wet and liquid, as it
had been before. Since, them, it cannot be hardened
by either a strong or a mild fire, its fire must
necessarily be of a degree set between mild and
strong. Because the whole Art consists in the right
regulation of the fire, with this fire, which must
have a carefully regulated strength, Mercury is
hardened and turned into earth, accomplishing
everything we desire. The reason why it must be
turned into earth is that when it is wet or liquid,

it is then so firm and dense that the flame cannot work in its form but the latter instead protects and covers it.

But when it has become earth, it is open and the fire controls it, penetrating through all its pores, altering its form, rendering it heterogeneous and separable, while it had previously been quite dense, covered, closed, homogeneous with the essential substance of Mercury, and of the same nature, also impossible to be separated from it.

The reason, however, why it was at first boiling and now no longer boils is that during such boiling part of its wateriness, which is the cause of its boiling, has been removed and consumed. If all the watery moisture were removed and consumed, it would no longer boil and get a beautiful azure-blue color.

And with this all the doubts that you have presented in your registered letter are dissolved and explained. The rules are right, provided the degree of the fire is thereby observed; but if a mistake in the degree of fire occurs, everything is spoiled and done wrongly.

Put up with my weakness. I cannot do much, but the little I know I am ready to impart to others,

especially to you and your friends. For the rest, I always desire to learn more, so that I might serve you great merit as you deserve. Your affectionate and devoted friend, Friederich Gualdus.
Venice, 2 December 1674.

Still Another Letter of Friederich Gualdus to a Priest His Good Friend

My Friend,

Your letter of the 25th inst. has been handed to me together with a basket of excellent cherries. You are always doing me one honor and favor after another, and I cannot but confess the truth, namely, that I have got in you another father, indeed more that a father after so many and long years, as my real father is dead. God be praised and Heaven thanked for such a great kindness.

Your desire to know the proportion of the water and the earth in their reduction can easily be satisfied, provided you first let totally go of all the utterances of the philosophers, because in this case they do not speak unanimously. For one wants to take 10 parts of water. The second speaks of nine, another again of seven parts. And Pontanus demands three parts. And thus there are still many others who have yet other opinions. Now we let go of all of them and consider what we have to do and what is necessary, in addition to the possibilities of nature.

I say, therefore, that to extract the fixed and dry core of the Mercury after taking such great pains and completing the Herculean labor, it is necessary to keep the Work dry at all times. Therefore little water, yes, very little water is required, especially at the beginning, so that the earth, which is the dry part, can always keep control over the water. With little no mistake can be made. All the philosophers rather assert and say: If you wish to turn earth into water, take three parts of water and one part of earth; but if you wish to turn water into earth, take three parts of earth and one part of water, etc. This is the right rule.

When now we turn water into earth, that is, reduce the water over the earth, and wish to make a dry mass, quite frozen and hard like a shining marble, we must give it very little water in one go and keep the mass dry, so that the dryness always has the upper hand. And in this way it can probably be done, although it will be a little difficult in the beginning and require some effort. But when the earth begins to give its water again, it will already be easier. Care must be taken, though, to remove that superfluous wetness, or urinal moisture every 8 or 15 days, as that raw part does not unite with the earth, because the earth only attracts the most cooked and tough part that it finds in the water and instead repels the raw part, so that it appears to be quite frozen.

Furthermore, to answer the question whether they have to be kneaded together or whether the water has to be poured at the bottom of the vessel and the earth put over it, I say that it is all the same. The difference only consists in the fire, for if the water is poured below and the earth added above, a stronger fire is required, that is, such a strong one that the water can sublimate and enter the earth; but not so strong that the water inundated the earth completely like the sea. Because, if a firm union and true marriage like Chibric [Kibrik] and Beya [Beja] is to occur, they must not be separated but joined, so that they can stand together in a continual union that can never be separated.

But if one wishes to knead them together and then puts them neat the fire, the latter must be fairly mild and small, because the moisture sublimates much more easily, while the earth stays open due to the rubbing (kneading). Although it is true that they unite much more firmly with less and longer heat.

Be it now one or the other manner, the degree of the fire must be carefully watched, so that the woman does not separate from the man some raw and watery part that disappears in a little smoke (or steam).

And this is what I can report in this matter. For, to tell the truth, the greatest part of the Herculean labor consists in finding the ash-colored earth, that is, in knowing how to separate the fixed part of our materia, which is quite volatile, and in this most of the alchemists blunder by wrongly considering the fixed body as something else. It is not so easy to go wrong in the rest, which is much surer to work, and there are not so many mistakes that can occur, just as you will experience and recognize that Sendivogius wrote the truth when he said: My center is highly fixed. Whoever has fixed that part will also fix its spirit, which has issued from it. With which I always remain, Your most affectionately devoted friend, Friederich Gualdus. Venice, 22 May 1678.

Conclusion of the Translator of the 1720 Edition

All the conclusions that can be drawn from the preceding report, however, are not so strong and certain that they could induce me to believe that our mortal life could be longer or shorter than the goal that has been set for it in the unchangeable degrees of Providence. True, I will admit that this is so very famous Stone can be found. Nor will I gainsay that those whom God had honored with his Grace really possess it, and that among those our Gualdus has to be especially reckoned. Nevertheless, it is far from being able to abolish Destiny or ward off those dangers against which neither medicines nor other help can do anything. For an unexpected fall that does not allow a man ot get up; the floods of the sea and the rivers which exhaust us quickly; buildings shaken and collapsing through earthquakes or other accidents, causing houses and residents to turn to ashes, all are such causes houses and residents to turn to ashes, all are such causes which Death has reserved for itself. I am sure that if death were a live body, he would laugh at our musings. Aside from this, I can also really believe that a good diet, as also the medicines that serve the preservation of the radical moisture and the natural heat, no less the medicine referred to by

Herr Amptman, may well keep us healthy in our life but not, as stated, prolong it. Accordingly, those are only defying death who can maintain their life more than others without some regulation or medicine.

Of this Gualdus the newspapers subsequently wrote a great deal, namely, that after leaving Venice, he was supposed to have stayed in Florence, Turin, and Paris, and finally at the Hague. Yes, after the papers reported last year that a man who had been considered an Adept at the Hague and for that reason moved away, had found his death in the Scheldt through the unfortunate overturning of his boat, was considered to be over 400 years old although he looked like a 40-year old, many now believe that it must have been this Friederich Gualdus, who was at last called from this world by God through such a death.

Finis

A Word from the Publisher

Thank you for purchasing this small work from The R.A.M.S. Library of Alchemy. During his lifetime, Hans Nintzel was dedicated to the identification, acquisition, study, retyping and, when necessary, translation of what he considered to be the most important known works on Alchemy. Hans was assisted by his sparse network of fellow Alchemists, all members of the Restorers of Alchemical Manuscripts Society (R.A.M.S.). I was an active member of R.A.M.S.

My goal is to publish all of the works originally made available through R.A.M.S. as photocopies. To facilitate this, I have chosen to have the books professionally printed. I also have a few titles that I intend to add to the original R.A.M.S. Library, selected by strict criteria established by Hans.

If you have a work on Alchemy that you believe should be a part of the R.A.M.S. Library, please contact me through R.A.M.S. Publishing Company.

Philip N. Wheeler

www.ingramcontent.com/pod-product-compliance
Lightning Source LLC
Chambersburg PA
CBHW080802180526
45168CB00006B/2300